# Lecture Notes in Energy

Volume 84

Lecture Notes in Energy (LNE) is a series that reports on new developments in the study of energy: from science and engineering to the analysis of energy policy. The series' scope includes but is not limited to, renewable and green energy, nuclear, fossil fuels and carbon capture, energy systems, energy storage and harvesting, batteries and fuel cells, power systems, energy efficiency, energy in buildings, energy policy, as well as energy-related topics in economics, management and transportation. Books published in LNE are original and timely and bridge between advanced textbooks and the forefront of research. Readers of LNE include postgraduate students and non-specialist researchers wishing to gain an accessible introduction to a field of research as well as professionals and researchers with a need for an up-to-date reference book on a well-defined topic. The series publishes single- and multi-authored volumes as well as advanced textbooks.

**Indexed in Scopus and EI Compendex** The Springer Energy board welcomes your book proposal. Please get in touch with the series via Anthony Doyle, Executive Editor, Springer (anthony.doyle@springer.com)

More information about this series at http://www.springer.com/series/8874

Paweł Ocłoń

# Renewable Energy Utilization Using Underground Energy Systems

Springer

Paweł Ocłoń
Department of Energy
Faculty of Environmental and Power Engineering
Cracow University of Technology
Kraków, Poland

ISSN 2195-1284 ISSN 2195-1292 (electronic)
Lecture Notes in Energy
ISBN 978-3-030-75227-9 ISBN 978-3-030-75228-6 (eBook)
https://doi.org/10.1007/978-3-030-75228-6



This Springer imprint is published by the registered company Springer Nature Switzerland AG
The registered company address is: Gewerbestrasse 11, 6330 Cham, Switzerland

# Introduction

This monograph presents a building heating system based solely on Renewable Energy Sources. Considering that the construction industry is responsible for about 38% of greenhouse gas emissions in Poland, it is necessary to look for innovative heating systems to reduce this effect. Most particulate pollutants are emitted by boilers fired with solid fuels. The energy standard of most residential buildings is much lower than the 2021 target level. One of the methods to reduce greenhouse gas emissions is advanced retrofit of buildings, which makes it possible to reduce energy consumption by about 30–80%. However, not every building can be modernized. For this reason, this work aims to present an advanced system of building heating which is scalable and can be applied to residential buildings demonstrating both high and low energy standards.

EU directives require the member states to increase the share of Renewable Energy Sources in the total energy mix (e.g. Directive 2009/28/EC, 2001/77/EC and 2003/30/EC). According to the European Energy Policy, in 2020 the EU has to achieve a 20% share of RES in the final consumption of energy. For Poland this limit was set at 15%. In 2007 it was also decided that by 2020 the annual energy consumption of the European Union must be reduced by 20%. Therefore, actions aimed at ensuring energy efficiency are increasingly recognized not only as a measure to ensure sustainable energy supply, reduce greenhouse gas emissions, increase security of supply and reduce expenditure on energy imports, but also as a measure to promote competitiveness of European economies.

As a consequence of actions aiming to reduce the consumption of energy in European economies, through the introduction of the EU Directive 2010/31/EU, it is assumed that from the end of 2020 all newly constructed single-family houses and larger residential and public buildings built in the EU member states should be characterized by low, almost zero demand for thermal energy. This means that low-energy houses, including passive buildings, will become increasingly popular. What is more, after 2020, only passive buildings will obtain a building permit with the EU support. Due to Polish climatic conditions, it is difficult to achieve energy parameters below the required 15 kWh/($m^2$/year), as stipulated by EU recommendations.

The main disadvantage of passive buildings is their price. The cost of constructing such a building in Poland can be up to 35% higher compared to a traditional house (in Germany, the cost of constructing a passive house is about 10% higher than of a traditional one). Also the materials, windows for example, are expensive and the price of an individual design of such a house, which should be adjusted to the plot, is high. Another disadvantage is the limited choice when it comes to the aesthetic qualities of the building itself and of the yard. Passive houses typically have a simple shape (no terraces, protruding rooms, multi hipped roofs, etc.), which does not always meet the taste and needs of the investor.

High energy efficiency of buildings can be achieved through advanced retrofit, which significantly reduces the cost of the building heating, or by using advanced zero-emission systems based on renewable energy sources. In order for residential buildings to comply with EU directives and meet the objectives of the climate and energy package, it is necessary to significantly increase the share of RES in building heating supply systems. Implementation of RES-based heating systems is particularly desirable in the case of older residential buildings with low energy standards. The implementation is particularly important to reduce GHG footprint, and provide primary energy savings for low energy standard residential buildings.

This monograph presents a system based on heat pumps, photovoltaic (PV) cells and solar collectors fitted with a sun-tracking mechanism. The system also includes a heat accumulation unit located in the ground. The system is scalable and can be applied to different types of residential buildings, but it is dedicated to multi-family housing. The system works very well in new buildings with underfloor heating, but it still has to be implemented and refined to achieve high energy efficiency for residential buildings with low energy standards. Within the project funded under the Horizon 2020 programme *RESHeat—Renewable energy system for residential building heating and electricity production*, leaded by the author of this monograph, the system will be implemented for multi-family buildings in Krakow, Limanowa and Palombara Sabrina (Italy) in two different climatic zones.

The monograph familiarizes the Reader with the principle of the system operation and presents a simplified mathematical model of the system key components, such as the PVT cells with a cooling system, the heat accumulation unit in the ground and the heat pump. It also presents the COP testing results of a ground source heat pump in a pilot system installed at the headquarters of the company *F.H.U. Urządzenia Chłodnicze Czamara.*

Furthermore, the monograph presents the optimization of high-voltage underground power cable systems, which are commonly used to link renewable energy sources with the existing electrical grid. A novel modified Jaya algorithm is applied to optimize material costs (the cost of cable backfill and cables) of the underground cable line, ensuring at the same time that the cable temperature is kept below 65 °C (design temperature). The cost optimization is important due to the following reasons:

(a) underground cable lines are prone to failure due to the cable conductor overheating;
(b) minimization of the costs related to the cable line and to the cable backfill material is necessary to ensure the highest performance of the system.

The monograph is composed of 10 chapters. Chapter 1 presents the state of renewable energy sources in Poland with a special emphasis on photovoltaic systems being the key element of the RESHeat project. Chapter 2 discusses different methods of heat accumulation in the ground. Chapter 3 presents building heating systems based on heat pumps assisted by solar installations. Chapter 4 presents the concept of the RESHeat system together with the results of the testing of the pilot system in Limanowa. Chapter 5 presents the mathematical model of the ground unit of heat accumulation. Calculations will be carried out to analyse the energy efficiency of the heat accumulation unit and the impact of the ground physical properties on the system efficiency. Chapter 6 presents optimization of the RESHeat system operating conditions using the Particle Swarm Optimization heuristic algorithm for two different diameters of the accumulation tank. Chapter 7 presents the mathematical model of the cooling system of PV cells together with experimental verification. Chapter 8 presents the RESHeat system economic analysis for two buildings with 20 and 50 apartments. A comparison is drawn between the RESHeat system and the systems currently used on the market of heating installations. Chapter 9 presents the RESHeat system advantages over other heating systems used in residential buildings at the moment. Chapter 10 presents the optimization of underground power cable systems using the modified Jaya algorithm. The summary of the book content and the conclusions drawn therefrom are included in the book backmatter.

I would like to thank the following people for their help and valuable content-related advice offered to me during the preparation of this monograph:

Petar Varbanov (*Brno University of Technology*), Andrea Vallatti (*La Sapienza University of Rome*), Marek Czamara (*F.H.U. Urządzenia Chłodnicze Czamara*), Mehmet Ali Yildirim (*F.H.U. Urządzenia Chłodnicze Czamara*), Jacek Drab (*F.H.U. Urządzenia Chłodnicze Czamara*), Wiesław Zima (*Department of Energy, Cracow University of Technology*), Piotr Cisek (*Department of Energy, Cracow University of Technology*), Filip Bartyzel (*Jacobs*), Maciej Ławryńczuk (*Warsaw University of Technology*). I would also like to thank my wife, Marzena, for her support and patience during the preparation of this monograph.

I would like to acknowledge the European Commission since the RESHeat is supported by European Union's Horizon 2020 research and innovation programme under grant agreement No 956255, project RESHeat (Renewable Energy System for Residential Building Heating and Electricity Production).

# Contents

# Symbols

| | |
|---|---|
| $A$ | Surface area, $m^2$ |
| $a$, $c$ | Coefficients of an exponential model of the unit cost factor (–) |
| $A_{PV}$ | Heat transfer surface area of photovoltaic cells, $m^2$ |
| $A_{sc}$ | Heat transfer surface area of solar collectors, $m^2$ |
| $b$ | The distance between the conductors' axes and the bottom of cable bedding layer (–) |
| $C$ | Length of the sample, m, $C = 1.02$ m |
| $c_p$ | Specific heat, J/(kg·K) |
| $C_{total}$ | Total material costs (cable costs plus thermal backfill material costs) per km of cable line, kEuro |
| $d$ | Diameter, m |
| $D$ | Tank (heat exchanger) diameter, m |
| $f$ | Frequency, Hz |
| $F_o$ | Objective (cost) function, mln € |
| $G_\beta$ | Direct solar radiation intensity, $W/m^2$ |
| $h$ | Cable burial depth, m |
| h | Enthalpy, kJ/kg |
| $h$ | Heat transfer coefficient, $W/(m^2k)$ |
| $H$ | Tank burial depth, m |
| $H_2$ | Ground domain depth, m |
| $I$ | Current loading, A |
| $J_o$ | Investment costs, € |
| $k$ | Thermal conductivity, W/(m K) |
| $k_s$, $k_p$ | Skin and proximity effect correction factors |
| kseg | Number of segments in the radiator (–) |
| $L$ | Tank length, m |
| $l$ | The spacing between two consecutive cables, m |
| $L_S$ | Energy consumed by the compressor, kJ |
| $m$ | Mass flow rate, kg/s |
| $MC_{total}$ | Overall material costs, € |
| n | Normal vector (–) |
| $n$ | Population size |

| | |
|---|---|
| $n$ | Time period under analysis |
| p | Pressure, bar |
| $p$ | The distance between the conductors' axes and the top of cable bedding layer, m |
| $PF$ | Penalty function (–) |
| $Q$ | Heat flow rate, W |
| $q$ | Heat flux, $W/m^2$ |
| $Q_b$ | Heat lost by the building, W |
| $Q_{dem}$ | Building cumulated demand for thermal energy, kWh |
| $Q_{ground}$ | Heat transfered between the tank and the ground, W |
| $Q_{HP}$ | Heat absorbed by the heat pump, W |
| $Q_K$ | Heat given up in the condenser, kJ |
| $Q_o$ | Heat absorbed in the evaporator, kJ |
| $Q_{solPV}$ | Heat supplied to the tank by PV cells, W |
| $Q_{solSC}$ | Heat supplied to the tank by solar collectors, W |
| $Q_v$ | Volumetric flow rate, $m^3/s$ |
| $q_v$ | Volumetric heat source, $W/m^3$ |
| $r$ | Discount rate (–) |
| $r, \phi$ | Polar coordinates, m |
| $r_1, r_2$ | Random numbers generated during each iteration of Jaya algorithm |
| $R_{e,AC}$ | Alternating current (AC) resistance, Ω/km |
| $R_{e,DC}$ | Direct current (DC) resistance, Ω/km |
| $R_{e,ref}$ | Reference cable conductor electrical resistance (calculated for 20 °C), Ω/km |
| $r_{in}$ | Cooling tube inner radius, mm |
| $r_{out}$ | Cooling tube outer radius, mm |
| s | Entropy, kJ/(kg K) |
| $s$ | The spacing between the right edge of the bedding layer and the side cable axis, m |
| $T$ | Temperature, °C |
| $T_1$ | Water temperature in the non-insulated tank, °C |
| $T_2$ | Water temperature in the insulated tank, °C |
| $T_{in}$ | Inlet temperature of the cooling system of PV panels, °C |
| $T_{load}$ | Temperature of the heat pump upper source, °C |
| $T_{out}$ | Outlet temperature from the cooling system of PV panels, °C |
| $T_r$ | Transmittance of the building, W/K |
| $T_{tank}$ | Water temperature in the tank, °C |
| $t_{year}$ | One-year period, s |
| $U_1$ | Water-ground overall heat transfer coefficient (non-insulated tank), $W/(m^2 K)$ |
| $U_2$ | Water-ground overall heat transfer coefficient (insulated tank), $W/(m^2 K)$ |
| $Uc$ | Unit costs, kEuro |
| $Ucf$ | Unit costs factor (–) |
| $U_g$ | Air-ground heat transfer coefficient, $W/(m^2 K)$ |

| | |
|---|---|
| $u_{in}$ | Fluid flow velocity in the tube, m/s |
| $v$ | Kinematic viscosity, m/s$^2$ |
| V | Tank volume, m$^3$ |
| w | Inertia coefficient in the Particle Swarm Method (–) |
| $W$ | Width, m |
| $W$ | Computational domain width, m |
| W, N, E, S, P | Nodes of control volumes |
| $x$ | Dryness factor (–) |
| **X** | Design variable vector |
| $x$, $y$ | Cartesian coordinates, m |
| $y_{EVA}$ | EVA encapsulation film thickness, mm |
| $y_{pv}$ | PV panel thickness, mm |
| $y_s$, $y_p$ | Skin and proximity coefficients |
| $y_{sz}$ | Glass layer thickness, mm |
| $\Delta x$, $\Delta y$, $\Delta z$ | Increment in direction $x$, $y$ and $z$, m |
| $\Delta t$ | Superheating value, °C |

## Greek Symbols

| | |
|---|---|
| $\alpha_{e,ref}$ | Temperature coefficient (–) |
| $\alpha_p$ | Thermal diffusivity, m$^2$/s |
| $\theta$ | Volume fraction of the material (–) |
| $\rho_w$ | Water density, kg/m$^3$ |
| $\rho_{20}$ | Specific electrical resistance of copper conductor in 20 °C, Ω m |
| $\eta_{sc}$ | Thermal efficiency of solar collectors (–) |
| $\eta_{PV}$ | Thermal efficiency of PVT panels (–) |
| $\phi(\tau)$ | Coefficient of momentary consumption of heat from the non-insulated tank |
| $\Delta\tau$ | Time step, s |
| $\xi$ | Friction factor for plain (smooth) tubes |
| $\mu$ | Dynamic viscosity coefficient, Pa s |
| $\Psi$ | Degree of heat demand coverage (–) |
| t | Time, s |

## Subscripts

| | |
|---|---|
| 1 | Non-insulated tank |
| 2 | Insulated tank |
| $g,s$ | Ground |
| $inl$ | Inlet |
| sz | Glass |
| $out$ | Outlet |

| | |
|---|---|
| *year* | Year |
| *tank* | Tank |
| EVA | EVA encapsulation film |
| PV | Photovoltaic cell |
| SC | Solar collector |
| HP | Heat pump |
| Load | Heat pump upper heat source |
| Source | Heat pump lower heat source |
| *al* | Aluminum |

## Criterial Numbers

| | |
|---|---|
| Nu | Nusselt number |
| Pr | Prandtl number |
| Re | Reynolds number |

## Abbreviations

| | |
|---|---|
| $CF$ | Cash flows, (€) |
| COP | Heat pump coefficient of performance (–) |
| $COP_{year}$ | Heat pump average annual coefficient of performance (–) |
| DPB | Discounted payback period, years |
| ERR | Refrigeration rate (–) |
| IRR | Internal rate of return, (€) |
| NPV | Net present value, (€) |
| NPVR | Net present value rate (–) |
| PVT | Photovoltaic thermal |
| SAGSHP | Solar-assisted ground source heat pump |
| SAHP | Solar-assisted heat pump |
| SPB | Simple payback period, years |
| UPCS | Underground power cable system |
| UTL | Underground transmission line |

# Chapter 1
# Renewable Energy Sources in Poland

The current legal regulations at the European Union level force a deep transformation of the Polish energy sector towards pro-environmental solutions (The European Green Deal 2019). Long-term goals of the EU energy policy include: achieving energy neutrality of the EU by 2050; maintaining economic growth, but without using non-renewable resources; maintaining sustainable development that will provide all EU regions with equitable development opportunities. The developed action plan aims to increase the effective use of raw material resources by moving towards a circular economy, as well as to restore biodiversity and reduce environmental pollution.

Achieving the EU's ambitious climate policy goals will require increased action from all sectors of the economy, including:

- investing in environmentally friendly technologies,
- supporting industry in innovation,
- introducing cleaner, cheaper and healthier forms of private and public transport,
- decarbonizing the energy sector,
- increasing energy efficiency of buildings,
- collaborating with international partners to improve global environmental standards.

It is therefore planned that in the years 2021–2027 the EU will allocate at least 100 billion euros to achieve these objectives. This is the support under the so-called Just Transition Fund to assist in the transition of national economies towards energy neutrality.

The next step towards EU energy neutrality is the framework programme for the years 2021–2030, which assumes (European Commission 2020):

- at least a 55% reduction of greenhouse gas emissions (compared to the 1990 level),
- at least a 32% share of renewable energy sources in the total coverage of the demand for energy compared to 1990 level,
- at least a 32.5% improvement in energy efficiency compared to 1990 level.



P. Ocłoń, *Renewable Energy Utilization Using Underground Energy Systems*, Lecture Notes in Energy 84, https://doi.org/10.1007/978-3-030-75228-6_1

For countries with a large potential of conventional power generation based on hard coal, which is the case in Poland, meeting the EU 2030 guidelines will be an extremely costly process. The decarbonization costs for Poland are higher than for other member states and meeting the EU guidelines may be difficult for the economy. However, Poland is taking steps towards gradual changes in the climate and energy sector. Among others, it is planned that funds will be allocated to the National Energy and Climate Plan for the years 2021–2030, in particular to such areas as (Luboińska 2020):

- nuclear power engineering,
- modernization of generating units in the power sector,
- development of transmission and distribution networks, including district heating networks,
- improvement of energy efficiency, including the construction industry, RES with the accompanying infrastructure, including dam structures enabling the installation of hydroelectric power plants,
- reduction of emissions from transport, including electro-mobility,
- storage of energy,
- development of the use of hydrogen fuels,
- increase in inter-system connections between EU countries,
- actions ensuring just transition.

Poland's National Power Grid, with the total installed power of over 41 GW, is in more than 70% based on coal-fired power plants. The oldest power units will be decommissioned in the coming years. According to the scenario of cumulated elimination of the existing generating units, as presented by the transmission grid operator until 2035, it will be necessary to shut down energy generation sources with the total capacity of 20 GW. This is due to the age and current state of the coal-fired power plants and to the planned implementation of proposals introducing new emissions standards (BAT – Best Available Techniques) (ERO 2019). For almost fifteen years, renewable energy sources have been developed actively in Poland, which is in line with both the climate policy of the European Union and global trends. As of the end of 2019, renewable energy sources accounted for 8.8 GW of power installed in the Polish Power System. Renewable energy sources are natural resources which are renewed in a short period of time. They include: water energy, solar energy, wind energy, geothermal energy and energy obtained from biomass processing. The state of the Polish RES sector, according to the Energy Regulatory Office (ERO 2020), is presented in Figs. 1.1 and 1.2.

Moreover, the Energy Regulatory Office presented the document "A report—summary of information on electricity generation from renewable energy sources in a small plant for the year 2018" (ERO, Summary of RES electricity generation 2019), which is summarized in Fig. 1.3.

In the document (Poland's Ministry of Energy, Poland's energy policy until 2040) i.e. "Polish Energy Policy (PEP) until 2040", prepared by the Ministry of Energy, the main objective of Poland's energy policy is formulated as follows: "The goal of the state energy policy is energy security, ensuring at the same time the competitiveness

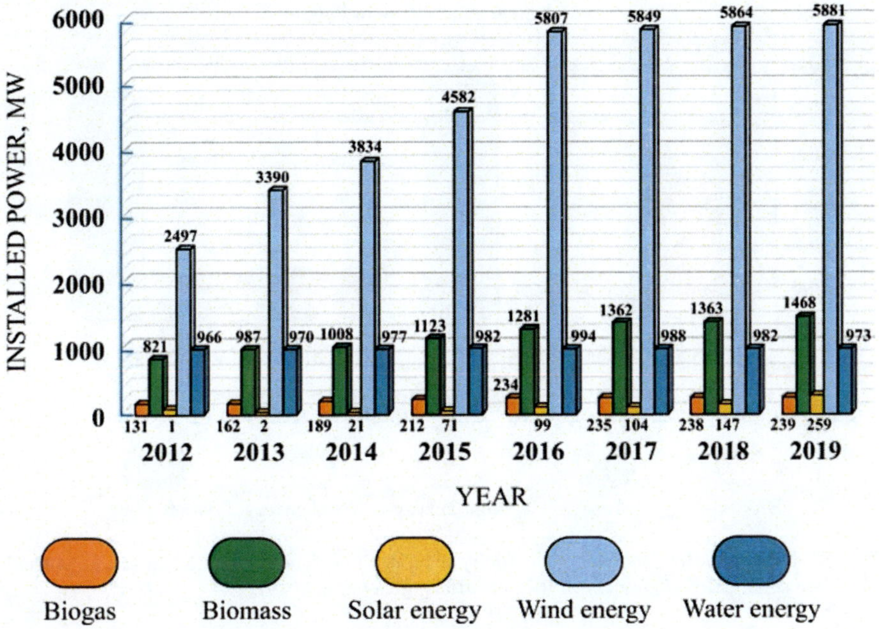


**Fig. 1.1** Installed power in Poland according to(ERO 2020)

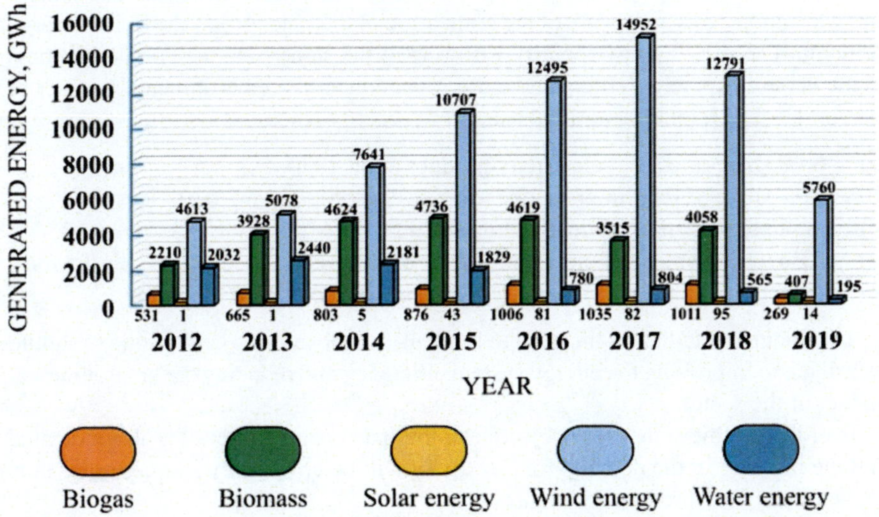


**Fig. 1.2** Electricity generated from RES in Poland in the years 2005–2019 confirmed by certificates of origin. (ERO 2020)

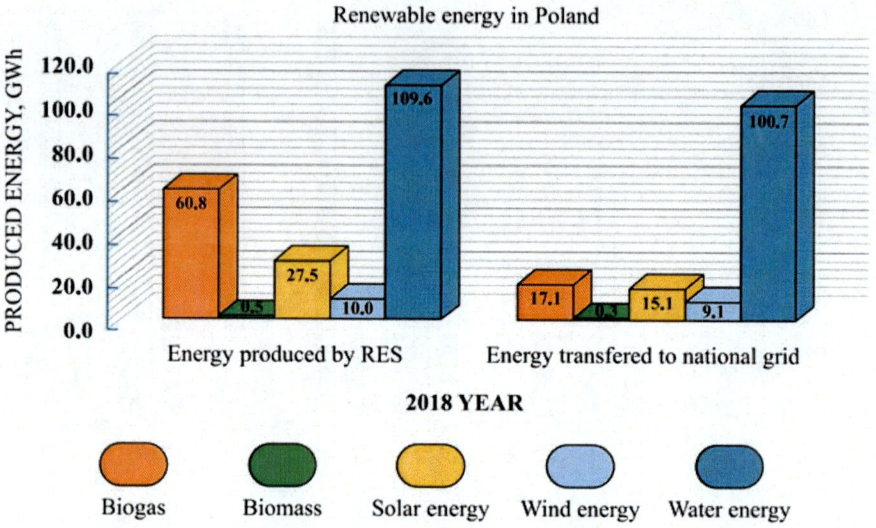


**Fig. 1.3** Data on the total amount of electricity generated in Poland from RES in a small plant and sold to the seller (ERO, Summary of RES electricity generation 2019)

of the economy, energy efficiency and a reduction in the environmental impact of the energy sector, making optimal use of own energy resources". The following indicators were adopted as a measure of the realization of the objective of the Polish energy policy until 2040: "Energy security, ensuring at the same time the competitiveness of the economy, energy efficiency and a reduction in the environmental impact of the energy sector, making optimal use of own energy resources":

(a) 60% share of coal in electricity production in 2030,
(b) 21% share of RES in gross final energy consumption in 2030,
(c) implementation of nuclear energy in 2033,
(d) 23% improvement in energy efficiency by 2030 compared to 2007 predictions,
(e) reduction of $CO_2$ emissions by 30% by 2030 (compared to 1990).

The document indicates the need to implement innovation – new solutions should contribute to improving the energy system efficiency and reducing the environmental impact of the sector.

In order to achieve the main objective of the state energy policy, it is also indicated that the increase in the demand for electricity will be covered by sources other than conventional coal-fired power plants.

The reduction in the emissions of pollutants from the energy sector will be achieved by using renewable energy sources. Further development of the use of energy from renewable sources will be considered as one of the instruments for reducing the environmental impact of energy generation. The share of RES in final energy consumption—approximate 21% in 2030—will result from cost-effectiveness and the possibility of balancing energy in the National Power System. The adopted target of the RES 21% share in gross final energy consumption in 2030 will translate into a 27% RES share in net electricity production. The key issue in achieving the electricity generation target will be the development of photovoltaics (especially from 2022 onwards) and offshore wind farms (the first offshore wind farm will be commissioned after 2025).

It is assumed that the share of RES in heating and cooling will grow at a moderate rate (about 1–1.3 percentage points per year), using the following energy sources: energy obtained from biomass (and heat from wastes), biogas energy, geothermal energy, heat pumps and solar energy (Poland's Ministry of Energy, Poland's energy policy until 2040).

The share of RES in electricity production in the next years to come will remain stable, although its dynamics may be accelerated after 2025 owing to the growing technical and economic maturity of individual technologies of energy generation (Poland's Ministry of Energy, Poland's energy policy until 2040, 2021)

It is estimated that in 2030 the share of RES in the energy sector will be in total about 27%. The use of RES will contribute to an increase in the RES share in the energy mix: solar energy (photovoltaics), offshore wind energy (to a very limited extent), biomass and biogas energy (mainly obtained in cogeneration), water energy (Poland's Ministry of Energy, Poland's energy policy until 2040, 2021)

It is expected that photovoltaic plants will reach their "production maturity" after 2022. The first offshore wind farms will be incorporated into the energy mix in Poland after 2025, after the expansion of transmission lines in the north of the country (Poland's Ministry of Energy, Poland's energy policy until 2040, 2021)

The unstable character of some renewable energy sources involves a number of issues in terms of the NPS flexibility and operational costs, compromising energy security and causing a rise in energy prices. To counteract these phenomena, local balancing will be required and RES will be used mainly to satisfy the needs of clusters and energy cooperatives. It is expected that after 2030 in Poland there will be about 300 local areas of sustainable energy development (Poland's Ministry of Energy, Poland's energy policy until 2040, 2021) The disadvantageous instability of RES will be compensated for by systems of energy storage. Considering the instability of supplies of energy obtained from RES, before RES can be incorporated into the power grid, it will soon become a prerequisite to ensure for them an adequate power reserve for periods of inactivity. Another option is energy storage.

Specific support mechanisms are still envisaged for RES, but these will depend on the type of source and its size, the nature of its operation—availability and controllability, the cost of energy generation and the extent to which local energy needs, including waste management, are met.

The basic forms of support are: priority access to the grid, auctions, the system of feed-in tariffs and surcharges, repayable assistance in the form of grants, guarantees of origin and technology-specific assistance mechanisms.

Apart from (Poland's Ministry of Energy, Poland's energy policy until 2040, 2021), there exists also the "National Energy and Climate Plan for the years 2021–2030" (NECP 2030). However, the documents lack a detailed analysis of electricity generation costs, their optimization and forecasts of energy prices. The results of

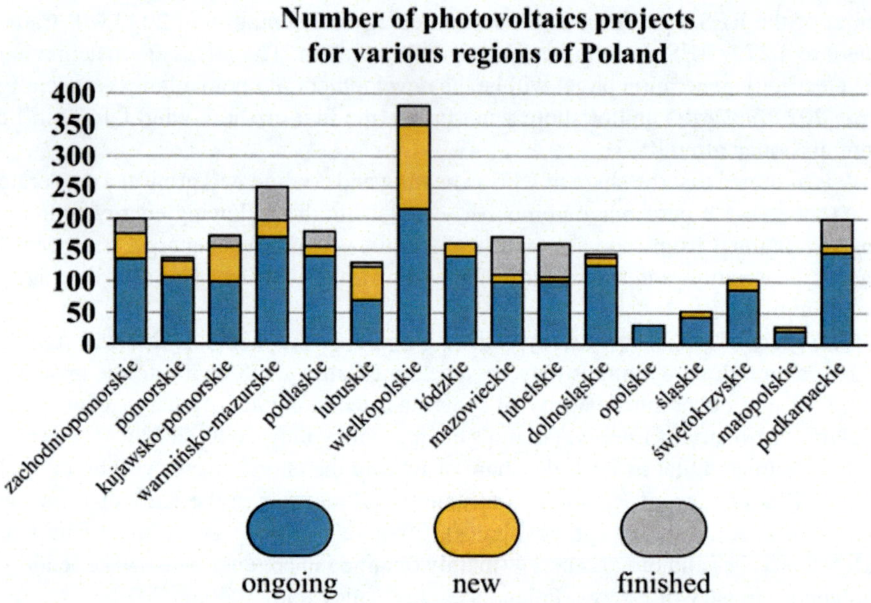


**Fig. 1.4** Number of photovoltaic projects in Poland according to the IRE: ongoing, new and finished (Institute for Renewable Energy 2020)

analyses conducted by the Polish Institute for Renewable Energy lead to a clear conclusion: an absolute increase in generation costs and tariffs is inevitable.

Further analyses confirmed that, even in the energy mix proposed in the NECP 2030, there is room for further optimization with regard to RES, whose introduction into the energy mix and full use mean lower energy costs. It is possible to reduce unit production costs by 3.7% in 2026 and by 8.5% in 2040 In the proposed energy mix, each additional megawatt of wind or photovoltaic energy reduces the cost of energy from the national power system.

Figure 1.4, developed by the IRE, shows for each voivodeship the number of photovoltaic projects (ongoing, new and finished). It is characteristic that new projects usually have the connection power of up to 1 MW and they obtained the environmental decision recently.

Analysing the state of photovoltaic projects in Poland, the IEO prepared a report that shows, among other things, the distribution of power of PV projects in the past several years (Figure 1.5).

Figure 1.6 confirms that the majority of new projects is in the range of 0.5 to 1 MW. The most optimistic scenario for the development of the PV sector is assumed by the PEP 2040 draft—the total power of PV installations in 2040 is expected to reach more than 20.2 GW (the NECP 2030 draft assumes the construction of 15.7 GW of PV installations in 2040). According to these assumptions (PEP), photovoltaics will account for about 25% of installed power in the year 2040.

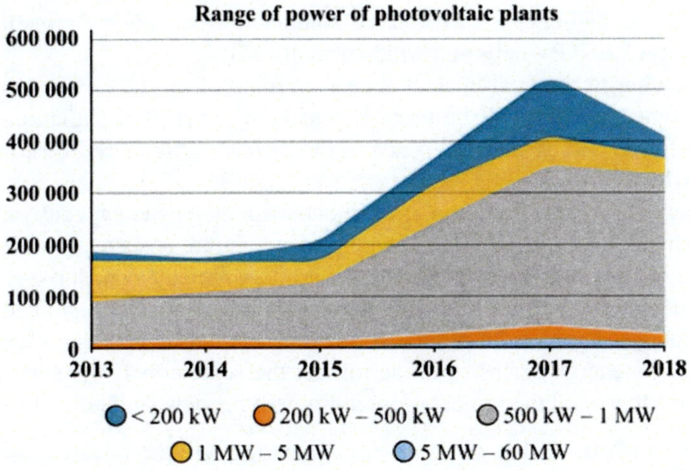


**Fig. 1.5** Power distribution of photovoltaic projects in Poland (Institute for Renewable Energy 2020)

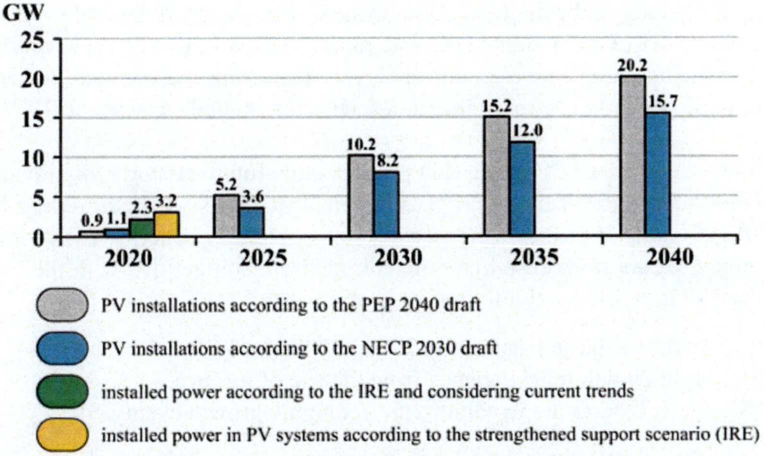


**Fig. 1.6** Market of PV installations in Poland (Institute for Renewable Energy 2020)

According to the IRE, the PV market will become the main area of investment in renewable energy.

The IRE's "Forecast of electricity costs in the perspective of 2030 and 2050—profitability of investment in fossil fuels and RES" shows the necessity of investing in RES and the related continuous improvement in the qualifications of people involved in the RES technology:

(1) The rising generation costs in the energy mix based on high-emission sources and, consequently, the higher prices of electricity, encourage the search for

alternative and cheaper RES-based solutions, especially by investing in the cheapest RES (PV cells and wind farms).

(2) According to the coal-nuclear scenario proposed by the government so far, electricity in 2030 will be more expensive by over 30% (inflation ignored) than today, and by 2050 the scale of the increase will exceed 60% compared to 2018.

(3) Dispersed in space and time, investments with the power capacity of 5–6 GW for small, distributed PV farms would reduce future costs by 1.5–2%.

(4) Renewable energy investments are the primary remedy against rising costs for businesses. Due to the current situation on the market, investments in RES reducing energy consumption in businesses are profitable under market conditions. As energy prices continue to rise, the scale effect (the source power) needed to ensure the project profitability will become smaller and smaller.

In the years 2005–2016, wind power engineering was the fastest-growing RES category in Poland, with an almost 70-fold rise. The record year for the increase in power was 2016, when 1225.38 MW were added. According to the data published by the Energy Regulatory Office, at the end of 2018 installed power of wind farms in Poland reached almost 5.9 GW, which corresponds to more than 14% of the total generating capacity of the national power system. The energy generated by the wind farms covered 7% of the national energy demand in 2018, and in 2017, a record year in terms of wind energy production, the level of meeting the national demand by onshore wind power was even higher at 8% (ERO, Installed capacity of RES 2019, 2020).

In 2020, the European Commission presented the future strategy for climate and energy measures, the so-called European Green Deal Strategy (The European Green Deal 2019). Its aim is to balance the member states in terms of energy. The economies of EU countries are to be resource-efficient, modern, competitive with the US and China, and should achieve the following goals:

(a) no greenhouse gas emissions by 2050;
(b) economic growth independence from the use of resources,
(c) EU member states are to achieve the economic growth evenly.

In order to achieve these targets, it is necessary to move towards the circular economy, which will enable increased resource efficiency (The European Green Deal 2019). Therefore, investment outlays are needed to improve energy efficiency, promote renewable energy sources and stimulate research on and development of RES.

It is anticipated that more than half of the new RES power plants commissioned in 2019 produce electricity that is cheaper than the product of the cheapest conventional power plants (European Commission 2020). Wind energy and solar energy have the biggest share in the RSE-based electricity market. For the ninth time in a row, solar energy reached the highest share (42.5%) in new investments using RES. The radiation of the sun is now most often used by photovoltaic installations converting solar energy to electricity. The cumulative annual growth rate of photovoltaic production

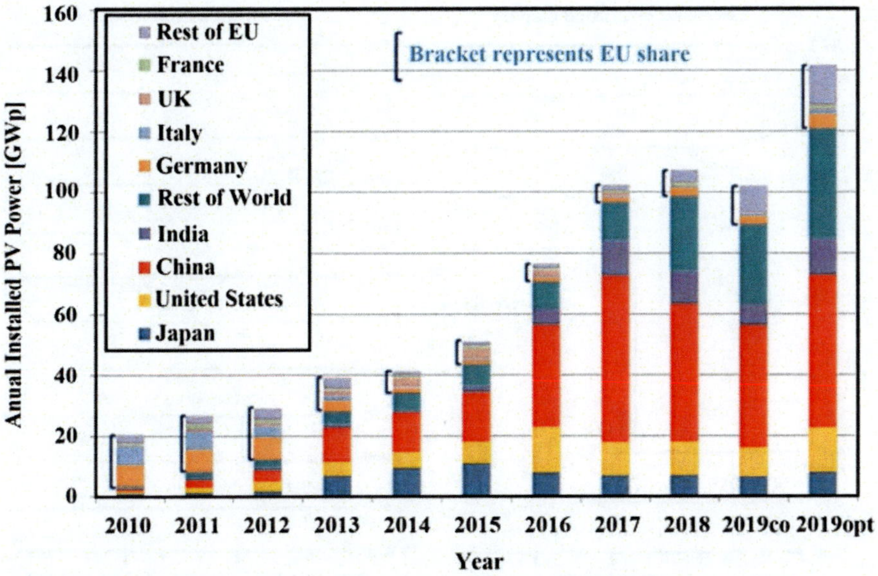


**Fig. 1.7** Worldwide capacity of photovoltaic installations in the years 2010–2019 (Taylor and Jäger-Waldau 2019)

for the last 15 years totals over 40%, so the photovoltaic industry is one of the fastest-growing in the world. The global annual power of newly installed PV plants grew from 29.5 GW in 2012 to 107 GW in 2018. The largest shares by country are listed below: China (44 GW), India (9 GW) and the United States (8 GW). The total power of photovoltaic installations in the world at the end of 2018 was 520 GW (Institute for Renewable Energy 2020). Figure 1.7 presents the global trend in the growth of power of photovoltaic installations installed worldwide.

The increased interest in the photovoltaic technology in recent years is mainly due to the decrease in the price of photovoltaic modules. In the case of solar energy, the levelized cost of electricity (LCOE) in 2010 was $0.371/kWh, while in 2018 it was $0.085/kWh, which means a drop by 77%.

Figure 1.8a shows the levelized cost of electricity over the years from 2010 to 2018 in the world. It was responsible for the decrease in the cost of PV panel production in 2018. In December 2018, the prices of PV modules in Europe ranged from $0.22/W, through $0.35/W, to $0.42/W for low-, medium-, and high-quality devices, respectively. Lower module costs reduce the cost of the entire photovoltaic system. The average start-up cost for a PV installation worldwide was $1210/kW in 2018, which means a decrease by 381% compared to 2010, when the price was $4,621/kW, as shown in Fig. 1.8b (Taylor and Jäger-Waldau 2019).

The rapid growth of interest in photovoltaic installations requires continuous refinement and development of photovoltaic technologies. Although the technology has been known for years, the conversion of solar energy to electricity using photovoltaic cells is inefficient. Photovoltaic cells absorb 80% of solar radiation, but the

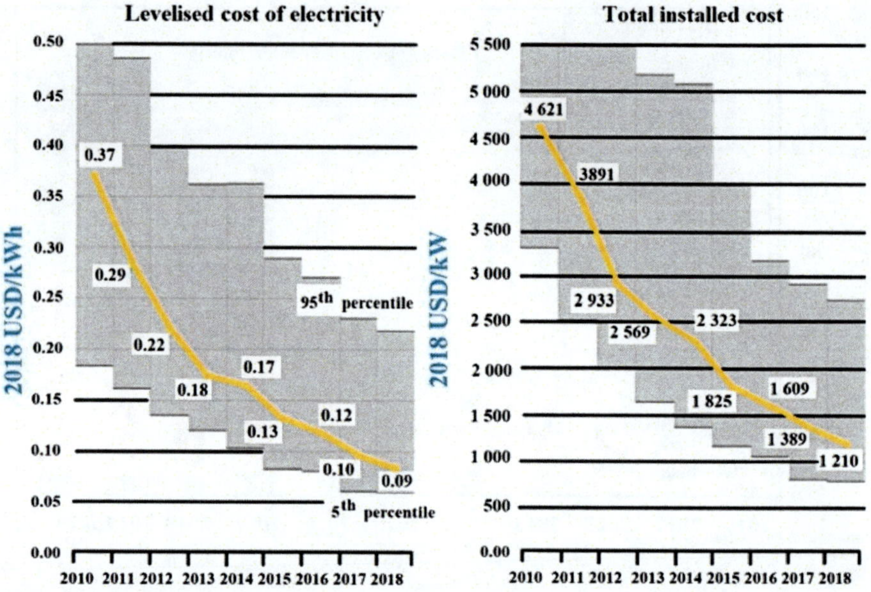


**Fig. 1.8** Average costs of a PV installation in the world over the years 2010–2018, **a** levelized cost of electricity, **b** average start-up cost of a 1 kWp PV installation (IRENA 2019)

efficiency of converting solar energy to electricity reaches the level of only 12 to 18%; in the case of monocrystalline cells it is up to 24%. It follows that some of the solar energy is lost irreversibly. In addition, the rest of the energy not utilized during photovoltaic conversion is converted to heat, which makes the temperature of the PV cells increase as high as above 50 °C (IRENA 2019).

The amount of solar energy reaching the Earth annually reaches $1.5 \times 10^9$ TWh, while the global energy consumption in 2019 was 158,839 TWh (Peng et al. 2017). This means that the sum of energy reaching the Earth is approx. 9.5 thousand times bigger than the global energy consumption. However, when it comes to the usable amount of solar radiation, it should be taken into account that it depends on the location of a given surface of the Earth and its position to the sun. A specific location is determined not only by latitude and longitude, but also by its external environment.

Figure 1.9a presents the solar energy resources in Europe. It can be seen that depending on the location of the country on the globe, a variation in solar radiation values is observed. Figure 1.10 shows the photovoltaic potential for European countries. Comparing the average radiation data between southern Sweden (1,010 W/m$^2$) and central Italy (1,420 W/m$^2$) for example, the difference can be up to 37%. The highest photovoltaic potential in Europe can be observed in Spain, where as much as 1800 kWh can be obtained from a 1 kW installation. Figures 1.11 and 1.12, respectively, present the map of solar radiation resources and of the photovoltaic power potential in Poland.

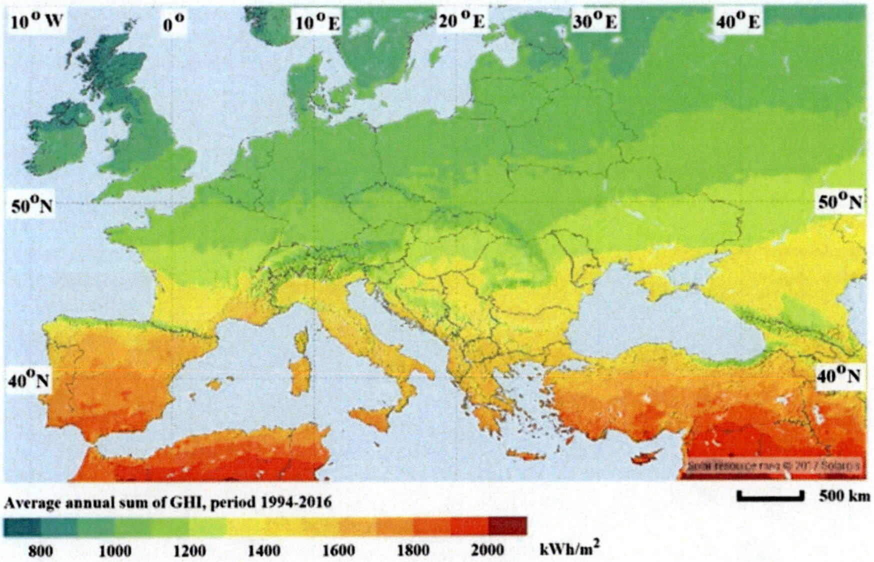


Fig. 1.9 Map of solar radiation resources in Europe (Solargis, Solar resource map of Europe 2019)

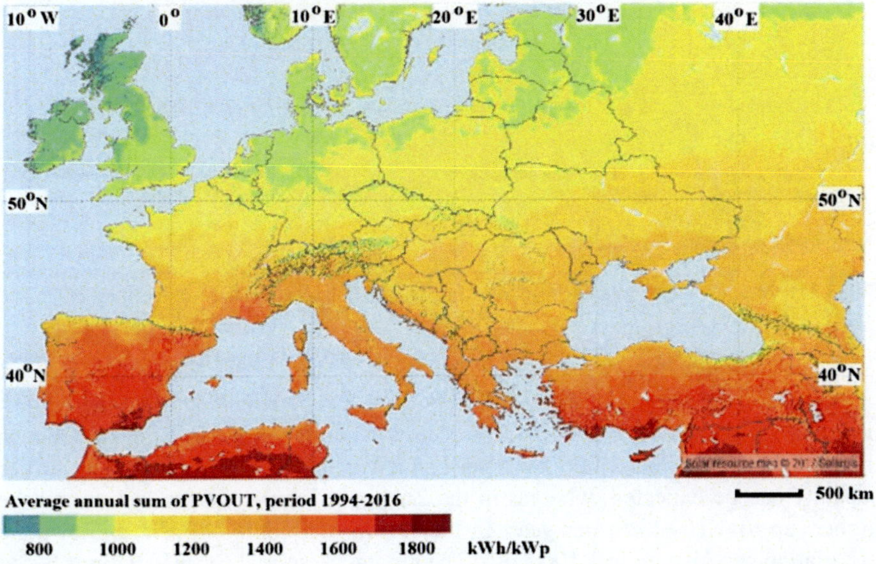


Fig. 1.10 Map of the photovoltaic power potential in Europe (Solargis, Solar resource map of Europe 2019)

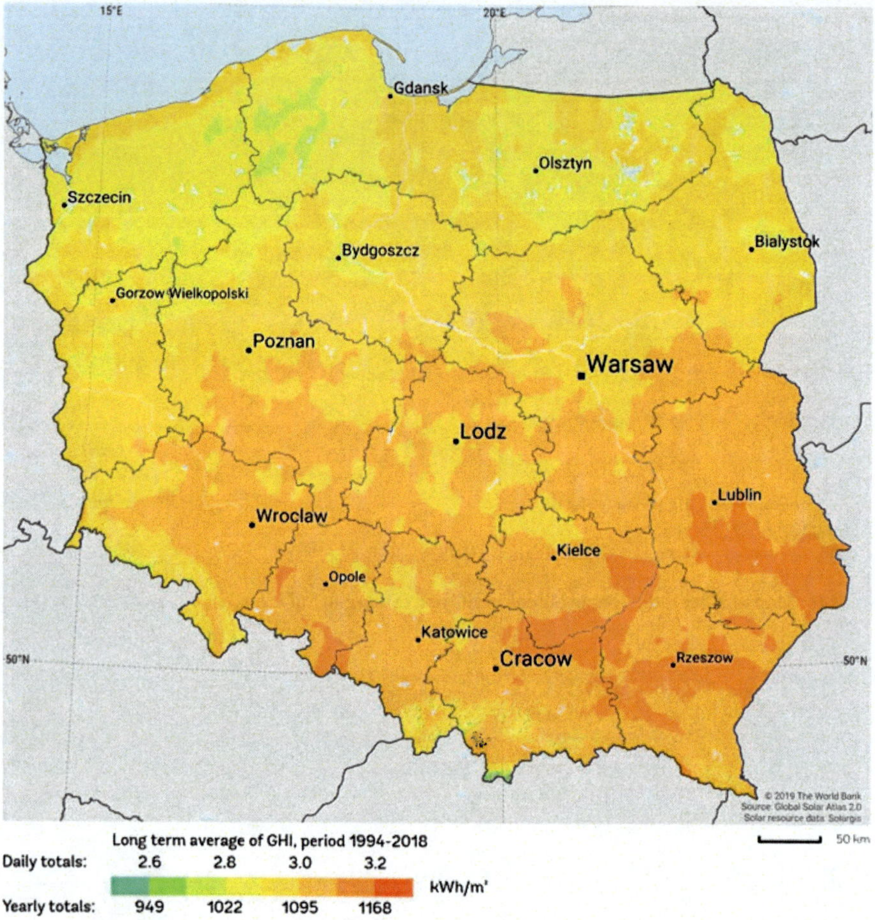


**Fig. 1.11.** Map of solar radiation resources in Poland (Solargis, Solar resource map of Poland 2019)

Analysing the two figures presented above, it can be concluded that the biggest potential for photovoltaic installations occurs in south-eastern Poland. In Krakow, the production of about 1,000 kWh per one kW of the photovoltaic system power capacity can be expected, whereas in the Zakopane region the production will be higher: up to 1,160 kWh per year. In the region of Lublin, a 1 kW photovoltaic installation can give up to 1,100 kWh. The greatest potential for installations based on PV cells and solar collectors is therefore in southern and south-eastern Poland.

This monograph presents a zero-emission heating system of a building using photovoltaic cells, solar collectors, heat pumps and energy storage in the ground. The proposed system will utilize thermal energy storage in the ground to improve the COP of the ground source heat pump.

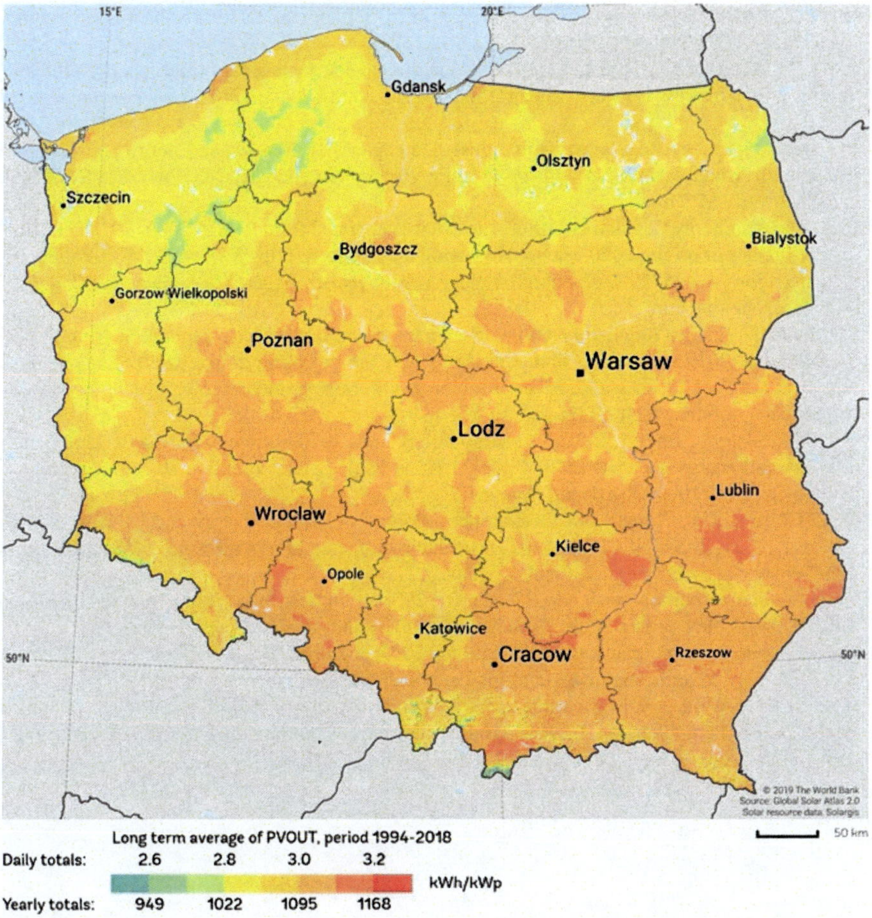


**Fig. 1.12** Map of the photovoltaic power potential in Poland (Solargis, Solar resource map of Poland 2019)

## References

Communication from the Commission to the European Parliament, the European Council, the Council, the European Economic and Social Committee and the Committee of the Regions, The European Green Deal, COM/2019/640

European Commission (2020), Climate and Energy Framework. Availabe online at: https://ec.europa.eu/clima/policies/strategies/2030

Luboińska U (2020) Greenhouse gas emissions. Selected issues concerning $CO_2$ emissions in Poland Thematic studies OT-683, Chancellery of the Senate, Office of analyses, documentation and correspondence, Warsaw 2020 (in polish)

Institute for Renewable Energy, PV Market in Poland 2020. Available online at: https://ieo.pl/en/pv-report

IRENA (2019) Renewable power generation costs in 2018. International Renewable Energy Agency, Abu Dhabi. ISBN 978-92-9260-126-3

Peng Z, Herfatmanesh MR, Liu Y (2017) Cooled solar PV panels for output energy efficiency optimisation. Energy Convers Manag 150:949–955. https://doi.org/10.1016/j.enconman.2017.07.007

ERO, Poland's Energy Regulatory Office, Installed capacity of Renewable Energy Sources, 2019 (in polish). Available online at: https://www.ure.gov.pl/pl/oze/potencjal-krajowy-oze/5753,Moc-zainstalowana-MW.html

ERO, Poland's Energy Regulatory Office, Electricity generated from RES in 2005–2019 confirmed by certificates of origin, 2020, (in polish). Available online at: https://www.ure.gov.pl/pl/oze/potencjal-krajowy-oze/5755,Ilosc-energii-elektrycznej-wytworzonej-z-OZE-w-latach-2005-2019-potwierdzonej-wy.html

ERO, Poland's Energy Regulatory Office, Summary on RES electricity generation from a small installations, 2019 . Available online at: https://bip.ure.gov.pl/bip/o-urzedzie/zadania-prezesa-ure/raport-oze-art-17-ustaw/3556,Raport-zbiorcze-informacje-dotyczace-wytwarzania-energii-elektrycznej-z-odnawial.html (in polish)

Poland's Ministry of Energy, Poland's energy policy until 2040—strategy for the development of the fuel and energy sector (PEP2040), 2021 (in polish). Available online at: https://monitorpolski.gov.pl/M2021000026401.pdf, Warsaw

Poland's Ministry of Climate and Environment, Poland's National Energy and Climate Plan for the years 2021–2030 (NECP PL), 2021. Available online at: https://www.gov.pl/web/klimat/national-energy-and-climate-plan-for-the-years-2021-2030.

Solargis—solar resource maps and GIS data for 200+ countries, Solar resource maps of Europe (2019). Available online at: https://solargis.com/maps-and-gis-data/download/europe

Solargis—solar resource maps and GIS data for 200+ countries, Solar resource maps of Poland (2019). https://solargis.com/maps-and-gis-data/download/poland

Taylor N, Jäger-Waldau A, Low carbon energy observatory photovoltaics technology development report 2018—Public Version, EUR 29916 EN , European Commission, Luxemburg, 2019. ISBN 978-92-76-12541-9. https://doi.org/10.2760/373667, JRC118300

# Chapter 2
# Storage of Thermal Energy in the Ground

Renewable energy sources are characterized by high instability of operation. For solar installations the generated amounts of electricity (photovoltaics and PVT panels) and heat (solar collectors and PVT cells) depend on insolation, which varies depending on the time of the day and on atmospheric conditions. This necessitates the storage of RES energy. As already mentioned, this monograph presents a system of heat pumps assisted by solar energy. For the system to be effective, thermal energy has to be accumulated in the ground.

This chapter briefly describes the thermal energy storage methods. A more detailed information can be found in (Banks 2012; Dincer and Rosen 2011; Demirel 2012; Mania and Kawa 2016).

One of the methods of thermal energy storage in amounts which are economically significant is to use heat and cool storage in the ground (Fig. 2.1). Using it allows to accumulate heat/cool underground for short or long term (seasonal). For example, the heat produced by solar collectors or photovoltaicthermal panels in summertime can be stored for a longer time period and reused in heating season. The selection of heat accumulator depends on soil properties, available space, as well as heating demand. Due to the economic reasons, the medium which is commonly used for thermal energy accumulation is water.

The methods of thermal energy storage can be divided based on (Sunliang 2010; Mania and Kawa 2016):

a) the accumulation period:

- Short Term Thermal Storage (STTS) for example in intermediate tanks, an electric boiler,
- Seasonal Thermal Energy Storage (STES) in the ground, in tanks; assisted by solar installations.

b) the storage location or medium:

- Aquifer Thermal Energy Storage (ATES),
- Borehole Thermal Energy Storage (BTES),



P. Ocłoń, *Renewable Energy Utilization Using Underground Energy Systems*, Lecture Notes in Energy 84, https://doi.org/10.1007/978-3-030-75228-6_2

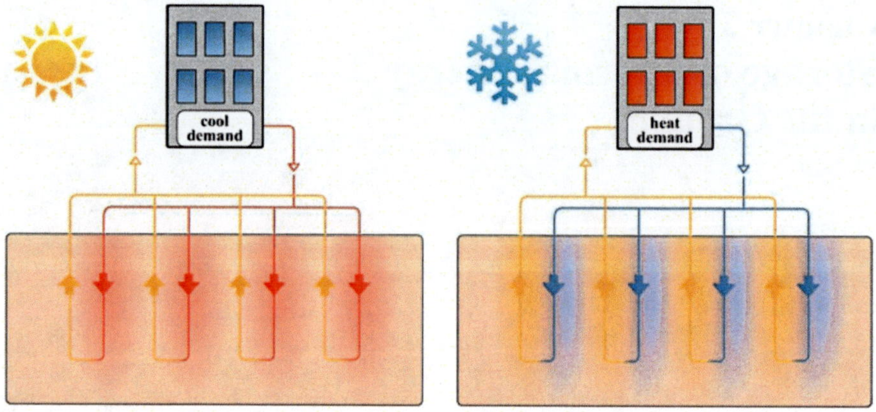


**Fig. 2.1** Heat and cool storage in the ground—redrawn based on (IF Technology BV)

in natural or man-made cavities, caverns, water-filled mining excavations – Cavity Thermal Energy Storage (CTES).

## 2.1 Storage of Thermal Energy in the Tank Thermal Energy Storage Method

The Tank Thermal Energy Storage (TTES) is a design solution using a water circular tanks (Fig. 2.2).The tank is using a reinforced concrete or steel structure, thermally insulated, closed from the top with a tight shell with the heating medium supply and return. After the burial in the ground the tank is covered with a soil to protect the tank individual layers. A discharge and recharge phase for heat storage (Heat TTES) is shown in (Fig. 2.3).

Hot water intake and inflow takes place in the upper part of the tank. Charging is done by introducing of hot water, and discharging is carried out by the removal of hot water. Such a clear separation of hot and cold water inflows allows thermal stratification in the tank, so that during unloading no mixing of layers takes place and high-temperature water is drawn at all times. In the same way, during the charging phase, low-temperature water flows into the heat source to increase the heat exchange rate.

The efficiency of this type of tanks is determined at the level of 60–80 kWh/m$^3$ (Mania and Kawa 2016).

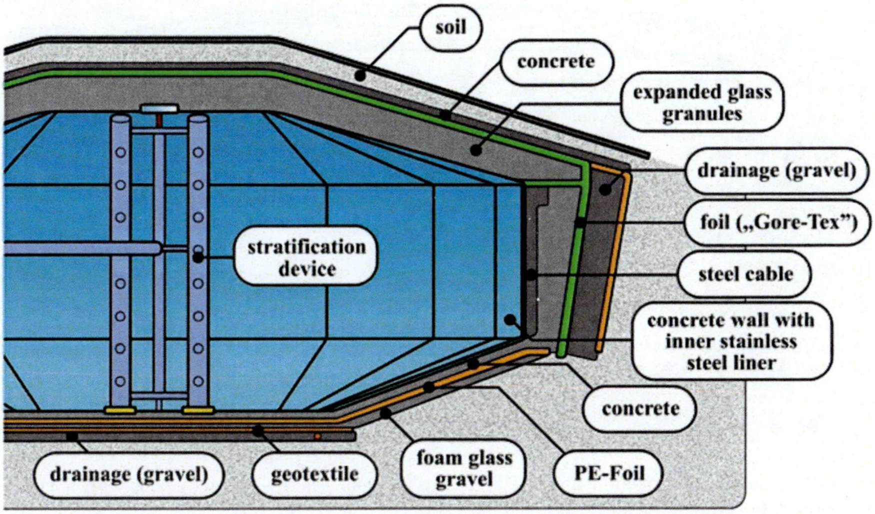


**Fig. 2.2** Cross-section of the TTES tank with structural details of individual layers—redrawn based on (Mania and Kawa 2016)

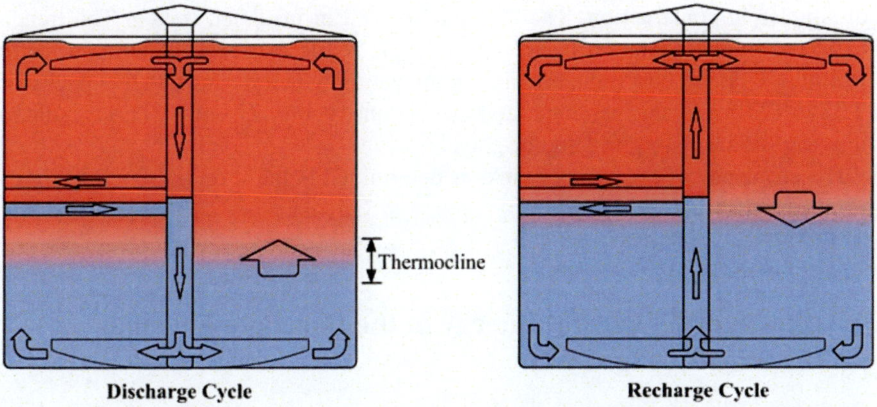


**Fig. 2.3** Discharge cycle and recharge cycle with a visible thin thermocline layer, where violent temperature changes occur

## 2.2 Storage of Thermal Energy in the Pit Thermal Energy Storage Method

Pit Thermal Energy Storage method uses large capacity water reservoirs with a slope of the pit side equal to approx. 1:2, depending on the ground conditions. The storages are not pressurized with a maximum temperature of 95°C. In this kind of heat storage solution, temperature stratification exists (Nielsen and Sørensen 2016).

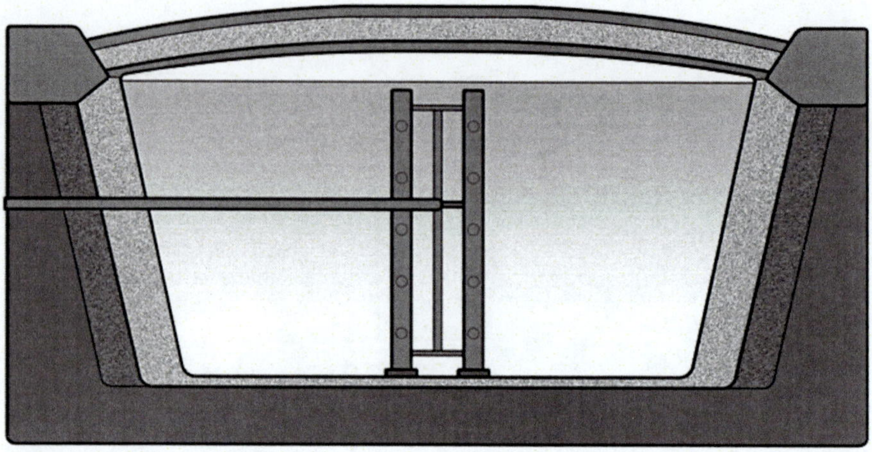

**Fig. 2.4** General cross-section of the PTES (Pit Thermal Energy Storage) tank with a visible insulation layer around the tank—redrawn based on (Mangold 2007)

The design facility can have any geometrical shape. A design is similar to Tank Thermal Energy Storage system (including a heating medium supply and return) but simpler in construction. The PTES is located in insulated trench. The trench is covered by an insulation layer, which is removable (Fig. 2.4). The main factor governing PTES is cost reduction and optimization which means larger capacity of PTES to TTES and cheaper material used for construction. Usually PTES is build in pit by adding insulating foil and/or gravel.

The efficiency of this type of tanks is determined at the level of 30–80 kWh/m$^3$ (Mania and Kawa 2016).

## 2.3 Storage of Thermal Energy in the Borehole Thermal Energy Storage Method

In the Borehole Thermal Energy Storage (BTES) the medium storing thermal energy is the ground (Dincer and Rosen 2011).The vertical probes are used to transfer the heat or cool energy. The heat transfer from the ground to vertical probes is based on both heat conduction and convection. The vertical probes are connected to each other in series or in parallel to achieve the effect of uniform recharge or discharge of the energy accumulator (Figs. 2.5 and 2.6).

The system consist of: casing, double U-pipe borehole heat exchanger, grouting pipe, thermally reduced grouting and thermally enhanced grouting. There exists also piping connections between vertical probes.

To enable an efficient operation of BTES systems, the ground water flows should not exist. This allows to avoid the dissipation of accumulated thermal energy. For

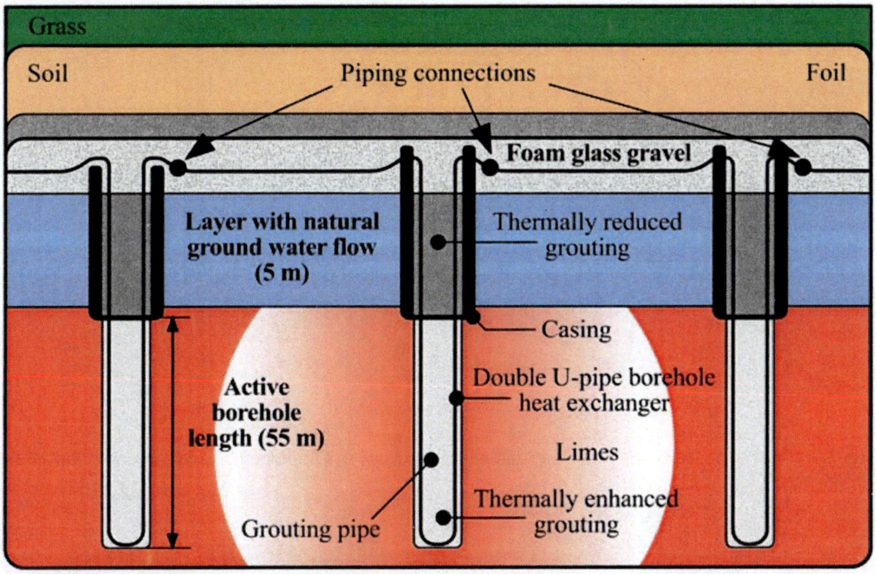


**Fig. 2.5** Borehole Thermal Energy Storage (BTES) with piping connections between vertical probes—redrawn based on (Mangold 2007)

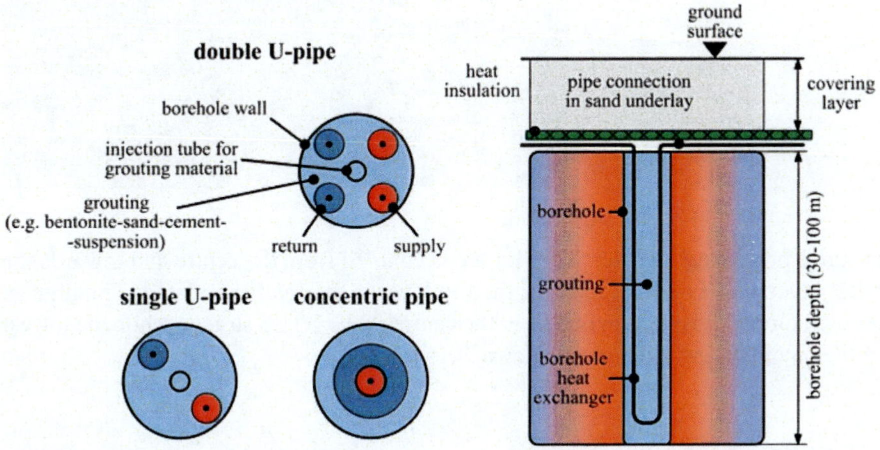


**Fig. 2.6** The design of vertical probe in BTES—redrawn based on (Mania and Kawa 2016)

an accurate design of BTES the numerical methods are mostly used to calculate the transient temperature and flow fields (Mania and Kawa 2016).

Figures 2.8 and 2.9, respectively, present the structure and construction of thermal energy storage using the BTES (Borehole Thermal Energy Storage) method in Okotoks in Canada. The vertical probes are located very close to each other to

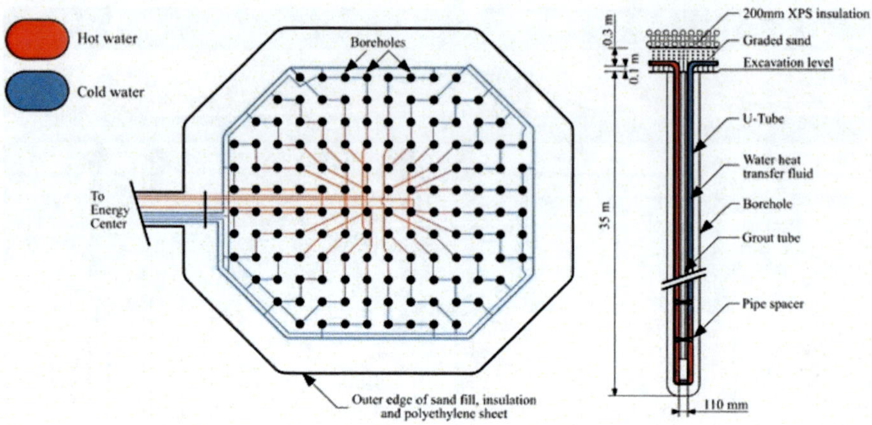


**Fig. 2.7** Design of the BTES-type thermal energy storage in Okotoks, Canada—redrawn based on (Drake Landing)

**Fig. 2.8** Photograph of the BTES-type thermal energy storage in Okotoks, Canada (Drake Landing)

ensure a better heat transfer. The storage is charged from the centre and towards the outer layers. To avoid heat losses, the top layer is thermally insulated. The highest temperature of the medium occurs in the centre of the BTES storage while the lowest in the outer coating (Mania and Kawa 2016).

## 2.4 Thermal Energy Storage in Reinforced Concrete Energy Piles

Another variant of the BTES-type energy storage system is based on reinforced concrete energy piles being a three-function element of the building structure (Fig. 2.9). The solution is very simple, and allows to increase the energy efficiency of the buildings by using the energy piles (Mania and Kawa 2016). It does not require additional costs related to excavations, as the most of the work is done during the

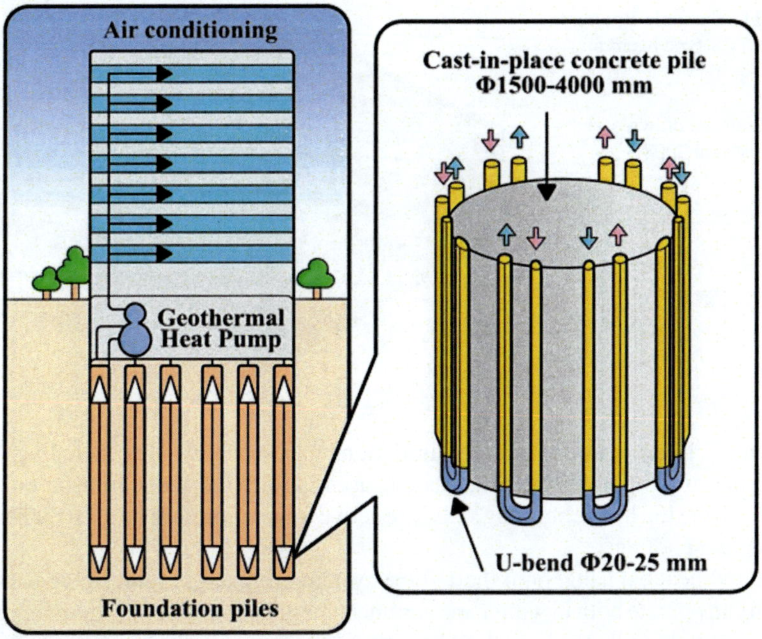


**Fig. 2.9** Cross-section of the BTES-type thermal energy storage with an energy pile embedded in the building foundation system—redrawn based on (Mania and Kawa 2016)

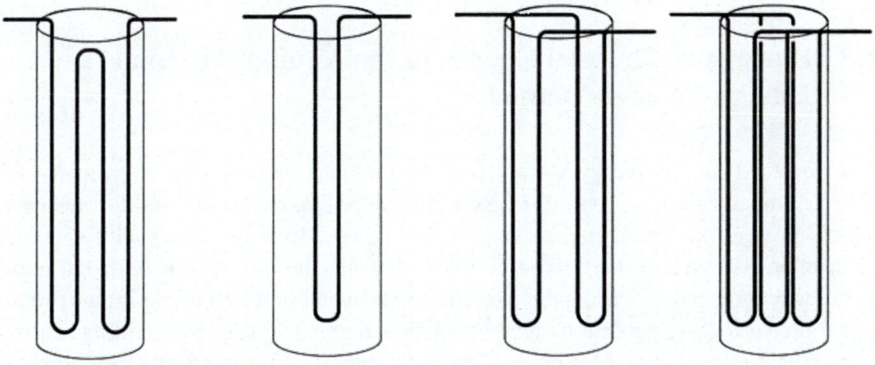

**Fig. 2.10** Different configuration types of energy piles—single, double and triple U-tube—redrawn based on (Song et al. 2021)

foundation construction. In order to not damage the foundation, the temperature of working fluid (circulating in the piles) shall not exceed 40°C in the summer, and be higher than 0°C in winter.

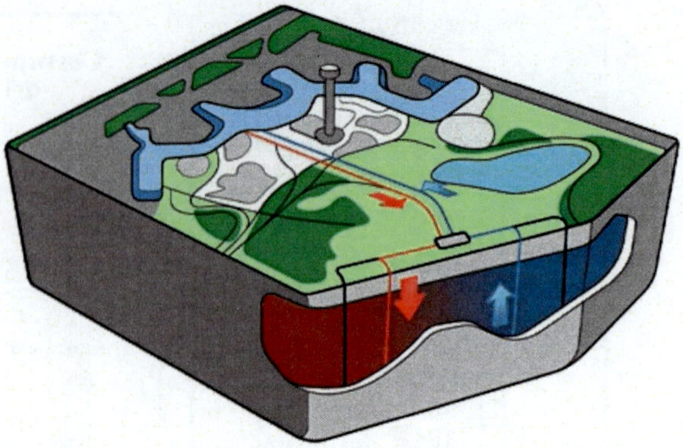

**Fig. 2.11** Example structure of the ATES-type reservoir on the aquifer at the Arlanda Airport in Stockholm—redrawn based on (Wigstrand 2008)

Figure 2.10 presents different configuration types of energy piles, including single, double and triple U-tube. The piles can be used to form the active foundations insulated from the top by the building slab. The active foundations can be used for building heating and cooling.

Ground heat exchangers in the U-tube system are fixed to reinforcing bars. The system can operate both in winter and summertime. The cool stored in the energy piles can be transferred through the building air-conditioning in the summer. Therefore, the cooling energy is provided. In winter time, the low-temperature heat stored in the piles can be transferred through the heat pumps for building heating.

## 2.5 Storage of Thermal Energy in the Aquifer Thermal Energy Storage Method

The Aquifer Thermal Energy Storage (ATES) are the systems where the groundwater is heated up and cooled down. ATES systems can be classified as open-loop systems, which means that circulating water pumped into the well is mixing with water in aquifer. In the system there exist couples of wells that are linked with the main groundwater reservoir. During the wintertime, the water in the reservoir is pumped up and cooled down in a heat exchanger, where the other side is used for building heating or as a heat source for a heat pump. The cold water is returned to the groundwater reservoir. For cooling purposes in the summer, the cold water is heated in the heat exchanger, while its other side is used for cooling (Nielsen and Sørensen, 2016).

Systems of this type (Figs. 2.11 and 2.12) use natural temperature capabilities of aquiferous layers and are a factor that increases the efficiency of heating and cooling systems. The chemical composition of such deposits must also be taken into account, as a high content of salt compounds or a high degree of iron content in the water can quickly lead to damage to the system. (Mania and Kawa 2016)

The ATES-type thermal energy storage system has been applied at Arlanda Airport in Stockholm (Figs. 2.11 and 2.13). The plant is expected to reduce the electricity

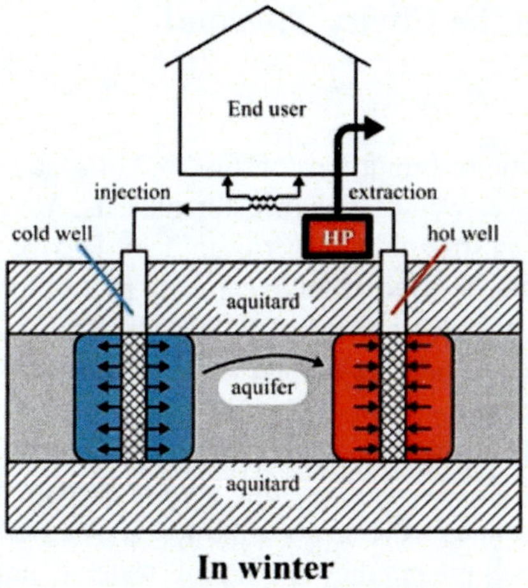


**Fig. 2.12** Example of ATES system (Klever 2018)

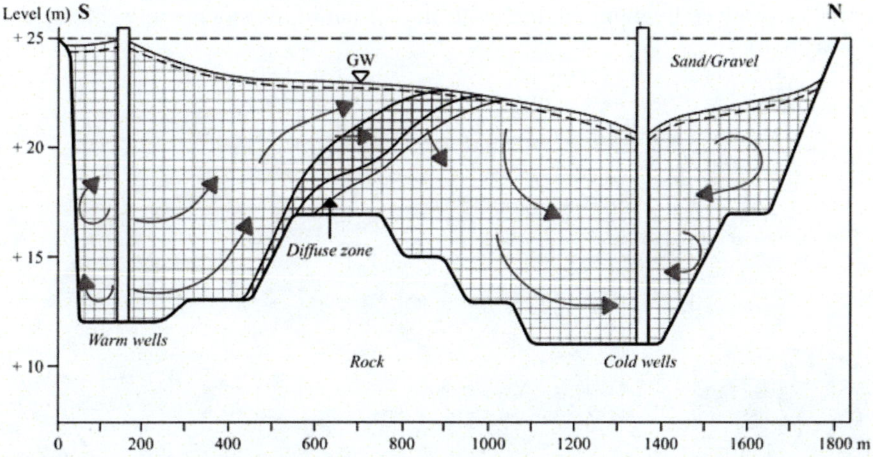


**Fig. 2.13** Cross-section of the ATES-type energy reservoir at the Arlanda Airport—redrawn based on (SWECO, 2007)

use by 4-5 GWh/year, and the district heating use by 10–15 GWh/year. Also, the $CO_2$-emmissions are expected to be decreased by 7,000 tones per year. The heating load capacity is of 8MW with a maximum ground water flow of 720 $m^3$/h (Wigstrand 2008).

## 2.6 Storage of Thermal Energy in the Cavity Thermal Energy Storage Method

The heat can be also stored in cavities. The major limitations of this kind of system are the technical adaptation and cavity sealing as well as high investment costs. Also, those systems application is limited by existing structures (caverns, mines). To avoid high investment costs, the existing caverns can be used as large capacity reservoirs. For example hard rock mines can be used as a natural reservoir because of their impermeable nature. The existing mines need to be adapted and some of the shafts and tunnels need to be sealed to create one continuous reservoir.

The Cavity Thermal Energy Storage (CTES) method is a design solution using existing underground excavations (Fig. 2.14). In this way, existing but not exploited caverns can be used for thermal energy storage and further heating purposes. Contrary to ATES, CTES is closed-loop system, which means that circulating water is not mixing with aquifer, which removes risk of contamination and damage done to system equipment by water impurities.

The CTES can be coupled with district heating systems. The thermal energy is stored or received by pumping the water into or out of the storage unit (Mania and Kawa 2016).

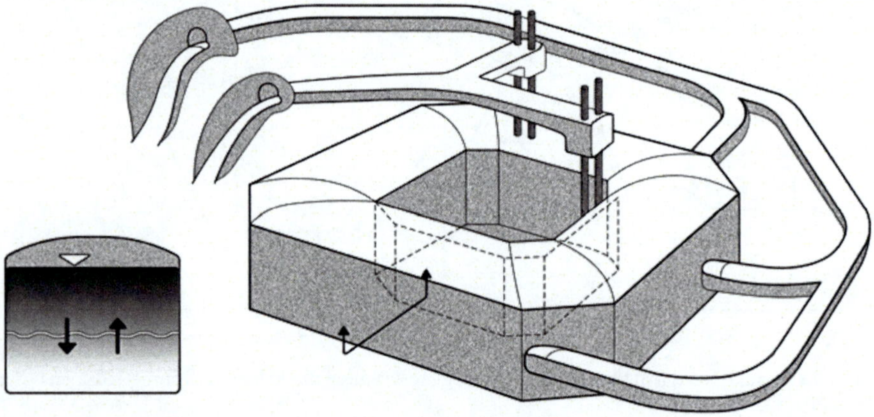

**Fig. 2.14** Example of Cavity Thermal Energy Storage (CTES) tank redrawn based on (Mania and Kawa 2016)

## References


Banks D (2012) An introduction to thermogeology: ground source heating and cooling. Wiley, The Atrium, Southern Gate, Chichester

Demirel Y (2012) energy production, conversion, storage, conservation, and coupling. Springer, Heidelberg

Dincer I, Rosen MA (2011) Thermal energy storage: systems and applications. Wiley, The Atrium, Southern Gate, Chichester

Drake Landing, Drake Landing Solar Community Okotoks. http://www.dlsc.ca/borehole.htm

IF Technology B.V., Borehole Thermal Energy Storage, company's marketing materials, available online at: https://www.iftechnology.com/borehole-thermal-energy-storage/

Klever Y (2018) The Netherlands leading in Aquifer Thermal Energy Storage, The Circonomist

Mangold D (2007) Seasonal Heat Storage–Pilot projects and experiences in Germany. Presentation at the PREHEAT Symposium at Intersolar

Mania T, Kawa J (2016) Engineering of thermal energy storage instalation. In: Mroźiński A (ed) Norway Projects Monograph. ISBN: 978-83-64423-37-6 (In Polish)

Nielsen JA, Sørensen PA (2016). Renewable district heating and cooling technologies with and without seasonal storage. In: Gerhard Stryi-Hipp (ed) Renewable heating and cooling, Woodhead Publishing, pp 197–220. ISBN 9781782422136

Pavlov G, Olesen BW (2009) Thermal energy in buildings—storage, concepts and applications. https://www-ttp.particle.uni-karlsruhe.de/GK/Workshop/blobel_maxlik.pdf

Song C, Li Y, Rajeh T, Ma L, Zhao J, Li W (2021) Application and development of ground source heat pump technology in China. Prot Control Mod Power Syst 6(17). https://doi.org/10.1186/s41601-021-00195-x

Sunliang C (2010) State of the art thermal energy storage solutions for buildings. MSc Thessis, University of Jyväskylä, Finland

SWECO (2007). Aquifer Storage at Långåsen, Arlanda Stockholm Airport. Permit Application. SWECO, Malmö, May 11, 2007 (in Swedish)

Wigstrand I (2008). The ATES project—a sustainable solution for Stockholm-Arlanda airport, accessed online: http://undergroundenergyllcnew.homestead.com/The_ATES_project___a_sustainable_solution_for_Stockholm-Arlanda_airport.pdf

# Chapter 3
# Solar-Assisted Heat Pumps

Energy efficiency and the use of Renewable Energy Sources are well-known key objectives of the European energy policy (European Commission 2011). The 2007 climate and energy package envisaged, taking into account the 2020 target and the 1990 situation, a 20% reduction in primary energy consumption in buildings, a 20% increase in energy production from renewable sources and a 20% reduction in greenhouse gas emissions. These targets were updated in relation to the year 2030 under the Climate and Energy Policy to reduce greenhouse gas emissions by 40% and increase the share of RES to 27%. Within these requirements, the heating/cooling section and the energy consumption of buildings (in Europe buildings consume about 40% of primary energy) receive much attention. In this regard, the reader should refer to the Energy Performance of Buildings Directive (EPBD) (European Commission 2002) and the revised version thereof (European Commission 2010). In general, in the heating/cooling sections and in buildings, the above-mentioned objectives can be achieved through actions on the demand side (modernization of building resources (D'Agostino et al. 2017a, 2017b)) and/or on the supply side (development of new technologies (European Commission 2016); the two sides are closely interrelated and their synergy has been explored within heat-mapping studies for Europe (Connolly et al. 2012, 2013). From the practical point of view, considering the side of the heat/cold supply, innovative hybrid energy systems based on RES are receiving more and more attention. Indeed, energy systems integrated with RES have caught keen interest of the engineering circles. For example, combinations of solar, photovoltaic (Vivas et al. 2018; Deshmukh and Deshmukh 2018) and thermoelectric systems (Zhu et. al. 2018) are investigated.

It is worth noting that the application of RES-based heat pumps is gaining interest also due to the broader framework of the heating sector. For example, studies on the thermal energy roadmap in Europe estimated that by 2050 the share of district heating could increase up to 50% of the total heat demand, with about 25–30% of thermal energy supplied by large electric heat pumps. This growth opens up the possibility of a significant use of alternative renewable heat sources.



P. Ocłoń, *Renewable Energy Utilization Using Underground Energy Systems*, Lecture Notes in Energy 84, https://doi.org/10.1007/978-3-030-75228-6_3

In the heating/cooling sector, heat pumps are among the technologies that hold the greatest interest and represent an important alternative to conventional systems. In this respect, multifunctional heat pumps can be used not only for heating and cooling, but also for the domestic hot water (DHW) supply, using different heat sources (e.g. ground, air or water). However, there is still considerable uncertainty as to their energy consumption in large-scale applications because field measurements have often shown lower efficiencies compared to laboratory-scale experiments (Caird et al. 2012).

When combining heat pumps with solar systems, air-to-water heat pumps are usually used. Indeed, the performance of heat pumps is highly dependent on the system operating conditions (e.g. the lower heat source temperature, the fluid outlet temperature, the thermal load). In particular, the coefficient of performance (COP) of air-source heat pumps depends on ambient temperature $T_{amb}$ (which is related to evaporation temperature $T_{eva}$). Therefore, during the winter season, when heating is needed and $T_{amb}$ is low, the efficiency of the system decreases due to the lower temperature of evaporation. Conversely, considering a reverse-cycle heat pump in the summer season, when cooling is needed and ambient temperature $T_{amb}$ is higher, the system efficiency decreases due to the higher temperature of condensation ($T_{cond}$). In addition, when the air-source heat pump operates in the heating mode, frost may form on the evaporator surface. This phenomenon occurs when the evaporator surface temperature is below both the freezing point of water and the dew point of moist air. In this situation, frost acts as a thermal insulator and additionally reduces the air flow surface area. In this way evaporation temperature $T_{eva}$ decreases the heat pump efficiency. For this reason, periodic defrosting is necessary. Currently, the most common method is reverse-cycle defrosting (Wenju et al. 2011): the refrigerant flow is reversed and hot steam from the compressor flows through an external coil, melting the frost. Due to the compressor wear if the room is not heated, the momentary efficiency of the heat pump decreases. For example, based on experimental testing, Madonna and Bazzocchi (2013) reported that the COP increased by 13% and 23% for 1 or 2 defrosting cycles per hour, respectively. This issue is of critical significance for air-source heat pumps, especially in a cold climate (Zhang et al. 2017; Bakirci and Yuksel 2011; Rehman et al. 2017) and when large-scale applications need to be considered. In this respect, coupling solar technologies with heat pumps, which are then referred to as a Solar-Assisted Heat Pump (SAHP), is a promising solution that makes it possible to overcome the above-mentioned limitations, reduce primary energy consumption, meet the targets set out in recent regulations and promote dissemination of air-source heat pumps in cold climates. The SAHP concept was first proposed by Sporn and Ambrose (1955). It was then investigated extensively to expand the application area (in hot and cold climates) and improve the system performance. In particular, different researchers tested multifunctional heat pumps for both room heating/cooling and domestic hot water production applications (Cai et al. 2016).

Heat pumps can be assisted using photovoltaic (PV) panels, solar collectors or hybrid photovoltaic thermal (PVT) panels. As shown in the literature review (Kamel et al. 2015), in practical applications heat pumps are integrated with thermal or PVT panels, either in a direct configuration (the working medium used to cool the panels is the heat pump lower source) or an indirect configuration. In the direct configuration, solar panels provide heat directly to the SAHP evaporator (Mohanraj et al. 2017). For example, in (Huang and Chyng 1999) a SAHP system for domestic hot water preparation is presented in (Amin and Hawlader 2015) where the authors investigated a desalination system based on a SAHP for domestic hot water production. In indirect SAHP systems, a heat exchanger is used to connect the solar system to the heat pump (Cai et al. 2016). SAHPs can be divided into three subcategories:

- parallel SAHP systems,
- serial SAHP systems,
- dual-source SAHP systems.

In parallel SAHP systems, the heat pump receives energy from the environment and the solar energy is supplied directly for room heating or domestic hot water production. In serial SAHP systems, solar energy is supplied to the heat pump evaporator, thereby raising the evaporation temperature (which increases the COP) and cooling the solar collectors (which increases the solar collector efficiency; (Chow 2010; Kalogirou 2004; Bombarda et al. 2016)). In dual-source SAHP systems, the evaporator can receive energy either from the environment (air) or from the solar collectors/cooling photovoltaic panels, depending on the system ambient and operating conditions.

Wang et al. (2011) performed laboratory-scale experimental studies on indirect dual-source ("air-source" and "water-source") SAHP evaporators for room heating and cooling and for water heating. An accumulation tank, connected to solar-thermal collectors, was used to supply heat to the "water" evaporator or to produce domestic hot water. The heat pump achieved the COP of 4, in the heating mode. Bridgeman and Harrison (2008) investigated the SAHP indirect performance in a serial configuration for domestic hot water preparation. The heat pump achieved a COP varying from 2.8 to 3.3 depending on the evaporator and the condenser temperatures. Loose et al. (2011) carried out pilot testing of different SAHP systems with different heat sources for room and water heating. The system used solar collectors and a geothermal heat pump with a ground heat exchanger: the collectors supplied the storage tank directly when sufficient solar radiation was available, otherwise the heat pump used waste heat from the solar collectors. Bakirci et al. (2011) investigated the performance of an indirect SAHP system intended for room heating. The solar collectors directly fed the storage tank connected to the heat pump evaporator. The testing results showed a COP in the range of 3.3–3.8. Bai et al. (2012) performed a theoretical study on an indirect hybrid SAHP system for domestic hot water production (based on PVT panels) using the TRNSYS modelling method. Seasonal performance for different climatic conditions (e.g., Hong Kong and different locations in France) and the average COP were simulated. The heat pump average COP was found at 4.9. Keliang et al. (2009) studied direct expansion of the SAHP using a numerical method (the distributed parameter method). Their results showed the importance of the variable-speed compressor in the supplied power modulation depending on operating conditions. Lwowie et al. (2015)

performed theoretical studies, using the parametric modelling method, on a dual-source auto-cascade SAHP system operating with a zeotropic R32/R290 mixture for domestic hot water production. They observed COPs in the range of 3.8–4. However, the system efficiency depends to a large extent on the composition of the mixture, and the solution requires further testing before it can be applied on a large scale. Jie et al. (2015) and Cai et al. (2016) investigated an indirect multifunctional SAHP system using experimental and numerical methods. The experimental study was used to validate the numerical model. The numerical model was then used to perform sensitivity analyses. The observed values of the COP varied in the range of 2.2–2.7. Based on the above-mentioned literature reviews and the ones performed earlier (Cai et al. 2016; Hepbasli and Kalinci 2009; Ozgener and Hepbasli 2007; Parida et al. 2011; Tian and Zhao 2013), it can be seen that there are no pilot studies or practical demonstrations of SAHP systems. For example, as stated by Cai et al. (2016): "very few theoretical studies on multifunctional solar-assisted heat pumps have been conducted, and the experimental verifications related thereto are still insufficient". Moreover, as noted by Croci et al. (2017), most studies focus on the application of SAHPs in northern European cold climates and no research has been carried out on a technical solution that could be used year-round in the Mediterranean area. In their work, Besagni et al. (2019) investigate a novel multifunctional solar-assisted dual-source heat pump (installed in a single family house in Milan). It is worth noting that the Milan area has a specific climate (cold and wet winters and hot summers). Table 3.1 summarizes the results of studies on different SAHP systems.

## 3.1 Solar-Assisted Ground Source Heat Pumps

The use of solar-assisted ground source heat pumps is presented in (Sun et al. 2020). In order to maintain energy efficiency at a high temperature and reduce energy losses of seasonal heat storage in solar-assisted ground source heat pumps (SAGSHPs), a novel SAGSHP system with heat cascading in boreholes was designed and the system field test was conducted. The ground heat exchangers were divided into two regions: the central region and the outer region. The former can maintain a high temperature (e.g., 45 °C) and the central region heat can be utilized directly without the need to use a heat pump. The annual average COP of the heat pump was obtained at 4.66.

Naranjo-Mendoza et al. (2019) describe studies on an experimental Solar-Assisted Ground Source Heat Pump (SAGSHP) system intended for residential heating. The system makes use of a series of shallow (1.5 m deep) vertical boreholes intended for seasonal heat storage in an underground "energy bank". The results show that after 19 months of operation, the system was able to demonstrate good performance in satisfying the building's heating needs during the winter period. Similarly, it is proved that solar energy delivered to the ground is useful not only to restore the soil thermal balance but also to store heat.

**Table 3.1** Summary of different SAHP systems

| SAHP system | System type | Main results | Literature | Year |
|---|---|---|---|---|
| Direct | | – COP close to 9<br>– Collector average efficiency 75% | Hawlader et al. (2001) | 2001 |
| Direct | | – COP: 3.32 in hybrid operation<br>– Improvement by 28.7% compared to an ordinary heat pump | Huang et al. (2005) | 2005 |
| Direct | | COP: 8 | Hawlader et al. (2001), Amin et al. (2013) | 2001, 2013 |
| Direct | | Energy-efficient solutions for (50–70 °C) DHW | Chaturvedi (2014) | 2014 |
| Direct | | – Average COP: 4.7<br>– Total efficiency: 69.7% | Zhou et al. (2016) | 2016 |
| Indirect | Parallel | COP improved by 11.2% | Liang et al. (2011) | 2011 |
| Indirect | Parallel | Annual COP: 3.2 | Izquierdo et al. (2015) | 2015 |
| Indirect | Serial | COP: 6.38 | Çağlar et al. (2012) | 2012 |
| Indirect | Serial | COP in the range of 3.0–3.4 | Bakirci et al. (2011b) | 2011 |
| Indirect | Serial | – Collector efficiency from 33 to 47%<br>– COP: 3.8 | Bakirci and Yuksel (2011a) | 2011 |
| Indirect | Dual | COP improved by 50% in the temperature of −7 °C | Liu et al. (2016) | 2016 |
| Indirect | Dual | – Electrical efficiency: 14.5%<br>– Average COP: 3.07 | Wang et al. (2015) | 2015 |
| Indirect | Dual | Average COP: 4.0 | Wang et al. (2011) | 2011 |
| Indirect | Dual | – COP: about 3.5<br>– Collector efficiency: 60% | Kaygusuz (1999) | 1999 |

A study on the energy efficiency and cost-effectiveness of a solar-assisted ground source heat pump (SAGSHP) system based on PVT cells is investigated in (Sakellariou et al. 2021) for the city of Thessaloniki (Greece). The SAGSHP system is to cover the demand for room heating and domestic hot water preparation in a low-rise residential building. A mathematical model developed in the TRNSYS program was used to perform a parametric analysis by varying the number of PVTs. It is worth noting that the two most important elements of the used model, the PVT cells and the geothermal heat exchanger, were verified against experimental data. Simulations were carried out and seven energy indicators were estimated based on their results to investigate the energy performance of the system from different perspectives. The authors found that the highest efficiency was achieved by the SAGSHP system with 16 PVT panels, which is able to cover 73% of the heating load and generate 1.22 times more electricity than it consumes. The electrical efficiency of the PVT panels did not change during the parametric analysis, and the maximum specific efficiency was estimated at 301.5 kWh per year.

Fine et al. (2018) studied a ground source heat pump system with a solar installation to eliminate the effect of the ground thermal imbalance and minimize the system operating costs. The model of the heat exchange from the ground with thermal mass is coupled with a time-marching model to analyse the system for different sizes of the solar system. Details related to this modelling technique are presented together with case studies to illustrate calculation results for three different buildings. It is shown that an increase in the solar system size can offset the thermal instability of the ground. However, bigger dimensions mean higher initial costs of the system. A cost-effectiveness analysis of the investment was then performed to determine the system operating cost as a function of the solar system size, and the optimal system size in relation to the system costs was found.

Nouri et al. (2019) present simulations of different configurations of solar collectors with a ground source heat pump system to meet the heating, cooling and hot water demand of a house in Tabriz, Iran. The configurations include: an indirect (parallel) system, an indirect (serial) system and a direct system (direct connection between the heat pump and the solar installation). The simulation is performed in the TRNSYS program for a 9 $m^2$ vacuum collector and three boreholes with the depth of 75 m. A comparison is made of power consumption and the COP of all configurations. The optimum configuration is the indirect system (parallel connection), which achieved the COP of 3.96. An economic analysis is also performed to compare the proposed system with the traditional solution, and the obtained payback period is about 13 years.

Verma and Murugesan (2017) investigated the performance of a Solar-Assisted Ground Source Heat Pump (SAGSHP) intended for thermal energy storage during the day and room heating at night. Experiments were carried out to evaluate the efficiency of thermal energy storage in the ground using a ground heat exchanger. Solar energy was converted from 9:00 a.m. to 5:00 p.m. and used for room heating during the night from 7:00 p.m. to 3:00 a.m. The experimental data were used to calculate the heat absorbed by the solar collector, the heat injected into the ground, the heat extracted from the ground and the COP of the system. The results indicate that the collector thermal power varied from 2.07 to 2.56 kW depending on the mass flow rate. An increase in the mass flow rate of the heat transfer medium in the collector and in the ground heat exchanger caused a 21% rise in the amount of heat injected into the ground. Ground recharge resulted in a 23% rise in the system COP for night-time room heating.

Dai et al. (2015) performed an experimental study on the effect of operating modes on the heating performance of a solar-assisted ground source heat pump system (SAGSHPS). The experiments were conducted in January and their aim was to investigate the SAGSHPS performance characteristics in different modes of the heat pump operation. The results show that solar heat can be used to accelerate ground regeneration with the heat pump module switched off, but the time of using solar energy to recharge the boreholes should be optimized depending on the water temperature in the solar storage tank to avoid unnecessary energy consumption by the circulation pump. In addition it is indicated that the solar tank of tap water has a beneficial effect on stable operation of the SAGSHPS.

Verma and Murugesan (2014) proposed a methodology for optimization of the solar collector surface area and the length of the ground heat exchanger. The paper aimed to achieve a higher COP of the solar-assisted ground source heat pump (SAGSHP) system using the Taguchi method. Key parameters, such as the solar collector surface area, the ground heat exchanger length, and the COP of the SAGSHP system were optimized to obtain a higher COP of the solar-assisted ground source heat pump (SAGSHP) using the Taguchi method. The seasonal COP was obtained at 4.23.

The first large-scale solar-assisted ground source heat pump (SAGSHP) intended to ensure heating and cooling of a village on the outskirts of Beijing was investigated experimentally and theoretically in (Huang et al. 2020). Long-term performance of the system with and without solar heating was monitored. A TRNSYS model of the SAGSHP system was developed and verified against measured data. Based on parametric studies and using the model, an optimal solution was proposed. It is demonstrated that adding a DHW system to the primary SAGSHP system will not only help ensure the ground thermal balance, but also reduce the primary energy consumption for hot water by 70%. The COP of the system increases by 9.4% from 2.42 to 2.65, and the annual total operating costs drop from CNY 892,000 to CNY 794,000 with only a slight increase in installation costs.

Emmi et al. (2015) studied the effect of integration of solar collectors and of the total length of boreholes on the energy efficiency of the ground source heat pump. A multi-storey residential building was analysed. The analysis covered the heat pump operation in a cold climate. An appropriate control strategy using the TRNSYS software was implemented to maximize the energy efficiency of the heat pump. The first step was the system sizing to meet the total heating load of the building. Without the assistance of solar collectors, the heat pump seasonal energy efficiency decreased by about 10% in each case under analysis.

The literature review makes it possible to identify the following research gaps:

(a) no studies on thermal and electrical efficiency of PVT panels in combination with water/water heat pump systems
(b) no studies on the ground thermal regeneration using waste heat from PVT cells and solar collectors
(c) no analysis of the impact of the ground regeneration on the heat pump COP or the heat pump post-regeneration energy efficiency

(d) no implementation of SAHP systems for multi-family housing in a moderate climate

(e) numerical simulations and optimization of solar-assisted heat pump systems are carried out mainly, but there are not enough field tests to evaluate the efficiency of such systems.

## References

Amin ZM, Hawlader MNA (2013) A review on solar assisted heat pump systems in Singapore. Renew Sustain Energ Rev 26:286–293

Amin ZM, Hawlader MNA (2015) Analysis of solar desalination system using heat pump. Renew Energ 74:116–123

European Commission (2016). An EU Strategy on Heating and Cooling. Luxembourg: Publications Office of the European Union

Bai Y, Chow TT, Ménézo C, Dupeyrat P (2012) Analysis of a hybrid PV/Thermal solar-assisted heat pump system for sports center water heating application. Int J Photoenerg

Bakirci K, Yuksel B (2011a) Experimental thermal performance of a solar source heat pump system for residential heating in cold climate region. Appl Therm Eng 31(8):1508–1518

Bakirci K, Ozyurt O, Comakli K et al (2011b) Energy analysis of a solar-ground source heat pump system with vertical closed-loop for heating applications. Energy 36(5):3224–3232

Besagni G, Croci L, Nesa R, Molinaroli L (2019) Field study of a novel solar-assisted dual-source multifunctional heat pump. Renew Energ 132:1185–1215

Bombarda P, Marcoberardino GD, Lucchini A, Leva S, Manzolini G, Molinaroli L, Pedranzini F, Simonetti R (2016) Thermal and electric performances of roll-bond flat plate applied to conventional PV modules for heat recovery. Appl Therm Eng 105:304–313

Bridgeman A, Harrison S (2008) Preliminary experimental evaluations of indirect solar assisted heat pump systems. In: Chez proceedings of the 3rd Canadian solar building conference, Fredericton

Çağlar A, Yamalı C (2012) Performance analysis of a solar-assisted heat pump with an evacuated tubular collector for domestic heating. Energ Build 54:22–28

Cai J, Ji J, Wang Y, Huang W (2016) Numerical simulation and experimental validation of indirect expansion solar-assisted multi-functional heat pump. Renew Energ 93:280–290

Caird S, Roy R, Potter S (2012) Domestic heat pumps in the UK: user behaviour, satisfaction and performance. Energ Effi 5:283–301

Chaturvedi SK, Gagrani VD, Abdel-Salam TM (2014) Solar-assisted heat pump—a sustainable system for low-temperature water heating applications. Energ Convers Manag 77:550–557

Chow TT (2010) A review on photovoltaic/thermal hybrid solar technology. Appl Energ 87:365–379

Connolly D, Mathiesen BV, Østergaard PA, Möller B, Nielsen S, Lund H, Trier D, Persson U, Nilsson D Werner S (2012) Heat Roadmap Europe 1: First Pre-study for the EU27

Connolly D, Mathiesen BV, Østergaard PA, Möller B, Nielsen S, Lund H, Persson U, Werner S, Grozinger J, Boermans T et al (2013) Heat Roadmap Europe 2: Second Pre-study for the EU27

Croci L, Molinaroli L, Quaglia P (2017) Dual source solar assisted heat pump model development, validation and comparison to conventional systems. Energ Procedia 140:408–422

D'Agostino D, Cuniberti B, Bertoldi P (2017) Data on European non-residential buildings. Data Brief 14:759–762

D'Agostino D, Cuniberti B, Bertoldi P (2017) Energy consumption and efficiency technology measures in European non-residential buildings. Energ Build 153:72–86

Dai L, Li S, Mu LD, Li X, Shang Y, Dong M (2015) Experimental performance analysis of a solar assisted ground source heat pump system under different heating operation modes. Appl Therm Eng 75:325–333

Deshmukh MK, Deshmukh SS (2008) Modeling of hybrid renewable energy systems. Renew Sustain Energ Rev 12:235–249
European Commission (2002). Directive 2002/91/EC on Energy Performance of Building. Luxembourg: Publications Office of the European Union
European Commission (2010). Directive 2010/31/EU Energy Performance of Building. Luxembourg: Publications Office of the European Union
Emmi G, Zarrella A, De Carli M, Galgaro A (2015) An analysis of solar assisted ground source heat pumps in cold climates. Energ Convers Manag 106:660–675
European Commission (2011). Energy Roadmap 2050, Luxembourg: Publications Office of the European Union
Fine JP, Nguyen HV, Friedman J, Leong WH, Dworkin SB (2018) A simplified ground thermal response model for analyzing solar-assisted ground source heat pump systems. Energ Convers Manag 165:276–290
Hawlader MNA, Chou SK, Ullah MZ (2001) The performance of a solar assisted heat pump water heating system. Appl Therm Eng 21(10):1049–1065
Hepbasli A, Kalinci Y (2009) A review of heat pump water heating systems. Renew Sustain Energ Rev 13:1211–1229
Huang BJ, Chyng JP (1999) Integral-type solar-assisted heat pump water heater. Renew Energ 16:731–734
Huang BJ, Lee JP, Chyng JP (2005) Heat-pipe enhanced solar-assisted heat pump water heater. Sol Energ 78(3):375–381
Huang J, Fan J, Furbo S (2020) Demonstration and optimization of a solar district heating system with ground source heat pumps. Sol Energ 202:171–189
Izquierdo M, de Agustín-Camacho P (2015) Solar heating by radiant floor: experimental results and emission reduction obtained with a micro photovoltaic–heat pump system. Appl Energ 147:297–307
Jie J, Jingyong C, Wenzhu H, Yan F (2015) Experimental study on the performance of solar-assisted multi-functional heat pump based on enthalpy difference lab with solar simulator. Renew Energ 75:381–388
Kalogirou SA (2004) Solar thermal collectors and applications. Prog Energ Combust Sci 30:231–295
Kamel RS, Fung AS, Dash PRH (2015) Solar systems and their integration with heat pumps: a review. 87:395–412
Kaygusuz K, Ayhan T (1999) Experimental and theoretical investigation of combined solar heat pump system for residential heating. Energ Convers Manag 40(13):1377–1396
Keliang L, Jie J, Tin-tai C, Gang P, Hanfeng H, Aiguo J, Jichun Y (2009) Performance study of a photovoltaic solar assisted heat pump with variable frequency compressor e a case study in Tibet. Renew Energ 34:2680–2687
Liang CH et al (2011) Study on the performance of a solar assisted air source heat pump system for building heating. Energ Build 43(9):2188–2196
Liu Y et al (2016) Performance of a solar air composite heat source heat pump system. Renew Energ 87:1053–1058
Loose A, Drück H, Hanke N, Thole F (2011) Field test and performance monitoring of combined solar thermal and heat pump systems. In: Chez ISES solar world congress, Kassel, Germany
Lv X, Yan G, Yu J (2015) Solar-assisted auto-cascade heat pump cycle with zeotropic mixture R32/R290 for small water heaters. Renew Energ 76:167–172
Madonna F, Bazzocchi F (2013) Annual performances of reversible air-to-water heat pumps in small residential buildings. Energ Build 65:299–309
Mohanraj M, Belyayev Y, Jayaraj S, Kaltayev A (2017) Research and developments on solar assisted compression heat pump systems e a comprehensive review (Part A: modeling and modifications). Renew Sustain Energ Rev

Naranjo-Mendoza C, Oyinlola MA, Wright AJ, Greenough RM (2019) Experimental study of a domestic solar-assisted ground source heat pump with seasonal underground thermal energy storage through shallow boreholes. Appl Therm Eng 162:114218
Nouri G, Noorollahi Y, Yousefi H (2019) Designing and optimization of solar assisted ground source heat pump system to supply heating, cooling and hot water demands. Geothermics 82:212–231
Ozgener O, Hepbasli A (2007) A review on the energy and exergy analysis of solar assisted heat pump systems. Renew Sustain Energ Rev 11:482–496
Parida B, Iniyan S, Goic R (2011) A review of solar photovoltaic technologies. Renew Sustain Energ Rev 15:1625–1636
Rehman H, Hirvonen J, Siren K (2017) A long-term performance analysis of three different configurations for community-sized solar heating systems in high latitudes. Renew Energ 113:479–493
Sakellariou EI, Axaopoulos PJ, Wright AJ (2021) Energy and economic evaluation of a solar assisted ground source heat pump system for a north Mediterranean city. Energ Build 231
Sporn P, Ambrose ER (1955) The heat pump and solar energy proceedings of the world symposium on applied. Sol Energ 11:1–5
Sun T, Yang L, Jin L, Luo Z, Zhang Y, Liu Y, Wang Z (2020) A novel solar-assisted ground-source heat pump (SAGSHP) with seasonal heat-storage and heat cascade utilization: field test and performance analysis. Solar Energ 201
Tian Y, Zhao CY (2013) A review of solar collectors and thermal energy storage in solar thermal applications. Appl Energ 104:538–553
Verma V, Murugesan K (2014) Optimization of solar assisted ground source heat pump system for space heating application by Taguchi method and utility concept. Energ Build 82:296–309
Verma V, Murugesan K (2017) Experimental study of solar energy storage and space heating using solar assisted ground source heat pump system for Indian climatic conditions. Energ Build 139:569–577
Vivas FJ, las Heras AD, Segura F, Andújar JM (2018) A review of energy management strategies for renewable hybrid energy systems with hydrogen backup. Renew Sustain Energ Rev 82:126–155
Wang Q, Liu Y, Liang G, Li J, Sun S, Chen G (2011) Development and experimental validation of a novel indirect-expansion solar-assisted multifunctional heat pump. 43:300–304
Wang Q et al (2011b) Development and experimental validation of a novel indirect expansion solar-assisted multifunctional heat pump. Energ Build 43(2):300–304
Wang G et al (2015) Experimental study on a novel PV/T air dual-heat-source composite heat pump water heating system. Energ Build 108:175–184
Wenju H, Yiqiang J, Minglu Q, Long N, Yang Y, Shiming D (2011) An experimental study on the operating performance of a novel reverse-cycle hot gas defrosting method for air source heat pumps. Appl Therm Eng 31:363–369
Zhang Y, Ma Q, Li B, Fan X, Fu Z (2017) Application of an air source heat pump (ASHP) for heating in Harbin, the coldest provincial capital of China. Energ Build 138:96–103
Zhou JZ et al (2016) Experimental investigation of a solar driven direct-expansion heat pump system employing the novel PV/micro-channels-evaporator modules. Appl Energ 178:484–495
Zhu T, Swaminathan-Gopalan K, Stephani K, Ertekin E (2018) Thermoelectric phonon-glass electron-crystal via ion beam patterning of silicon. Phys Rev B 97:174201

# Chapter 4
# Zero-Emission Building Heating System Using Thermal Energy Accumulation in the Ground

## 4.1 Concept of the System

Due to the need to increase the share of renewable energy sources for the purposes of residential heating, a system is proposed using PVT cells and vacuum collectors with a sun-tracking system coupled with a ground source heat pump and a heat accumulation unit in the ground. The system was implemented by the F.H.U Urządzenia Chłodnicze Czamara company under the SOPSAR project (financed by the National Centre for Research and Development—fast track program for small and medium enterprises). Department of Energy of the Cracow University of Technology is a partner in the project, responsible for the development of the program for the system thermal calculations and the program for numerical simulations of the system operation. Within the RESHeat project (funded from the H2020 framework), the Department of Energy will develop tools for modelling the SOPSAR system and assist in the implementation of the system in residential buildings (Poland and Italy).

The proposed RESHeat system enables combined cooling, heat and power (CCHP) generation, including seasonal energy storage in the ground. Waste heat from PVT panels and heat from solar collectors equipped with a sun-tracking system are stored. The general concept of the energy cycle is shown in Fig. 4.1.

The RESHeat system enables:

- utilization of solar heat as the basic source of energy,
- production of heat and electricity using photovoltaic thermal (PVT) panels,
- seasonal energy storage in the ground,
- supply of heat and cool to the building using a heat pump,
- supply of electricity to the building through a system of PV panels,
- utilization of waste heat from intelligent solar collectors or PVT panels for the ground regeneration,
- maintaining a high COP of the heat pump over consecutive heating seasons using ground regeneration,
- accumulation of waste heat from different sources (e.g., heat recovery from air-conditioning or from technological processes).



P. Ocłoń, *Renewable Energy Utilization Using Underground Energy Systems*, Lecture Notes in Energy 84, https://doi.org/10.1007/978-3-030-75228-6_4

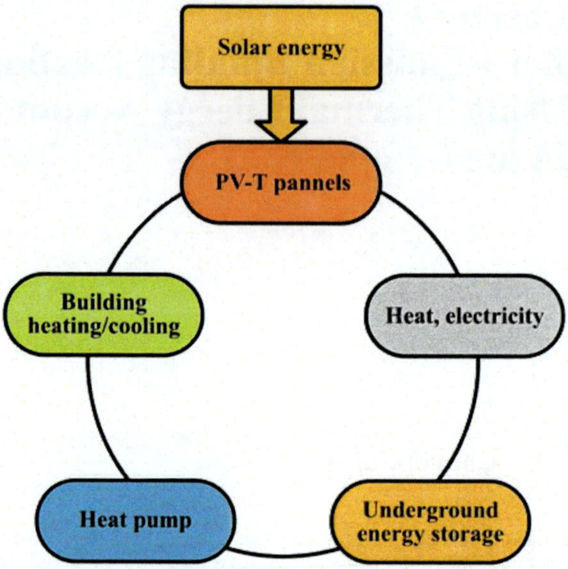


**Fig. 4.1** Concept of the RESHeat system

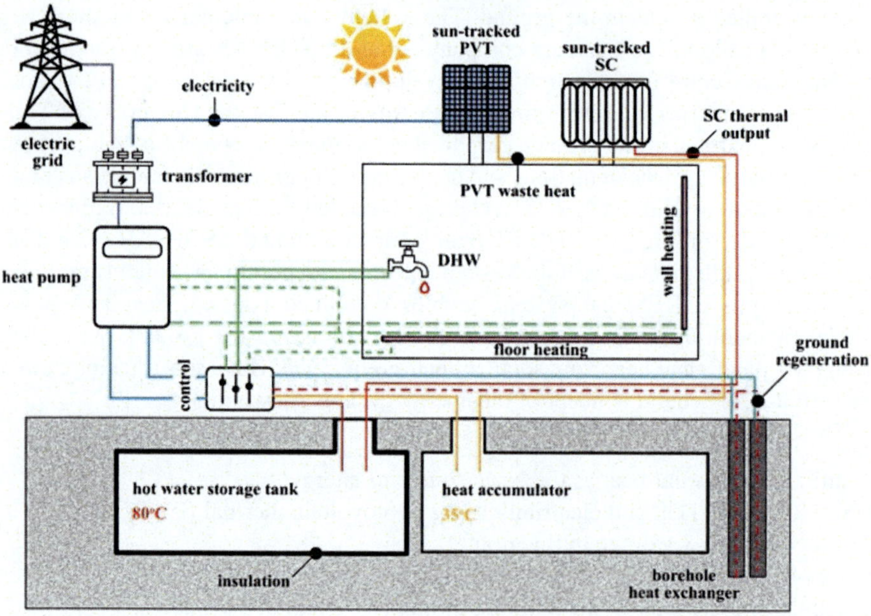


**Fig. 4.2** Schematic diagram of the RESHeat heating system

The proposed RESHeat system (Fig. 4.2) is a low-temperature (~35 °C) heating system intended for buildings. The system is based on a heat pump, waste heat recovery from the cooling of PV panels and underground thermal energy storage.

The main innovations of the proposed system are described below.

(a) The combination of waste heat recovery from PVT cells and underground energy storage makes it possible to achieve a high COP of the heat pump with an annual average value higher than 4. The maximum seasonal COP of typical water-to-water heat pumps totals 3.5. Such a high COP is due to efficient underground energy storage, making it possible to reduce the electricity consumption of the heat pump compressor by 20%.
(b) Development of the technology of high-efficiency ground regeneration through the heat transfer from an underground accumulator to the ground and from boreholes to the ground. Due to that, the heat pump COP does not decrease in consecutive years. In fact, this is the most significant innovation of the RESHeat solution, as it makes it possible to keep a constant COP over the years of the heat pump operation. No solutions are available on the market to actively regenerate the ground and keep the heat pump COP constant—the coefficient drops from one year to another, decreasing the pump efficiency.
(c) Cooled PVT panels with an intelligent sun-tracking system to achieve an 18% conversion of solar energy to electricity and a 60% conversion of solar energy to heat to maximize the efficiency of the RES system.

The non-insulated underground heat accumulator can alternatively be charged by solar collectors (only if there is an excess of thermal energy in the insulated tank). The heat absorbed by the ground can be collected from the accumulator surface and transferred to another location with ground water. Therefore, losses should be prevented by insulating the domestic hot water tank. The insulated tank can also be used as the peak heat source in winter, when the outdoor temperature is very low. Also in the autumn–winter period, when the output temperature from the solar collectors is below 40 °C, waste heat is used to regenerate the ground through vertical heat exchangers.

The idea of the system is as follows: the waste heat recovered from the cooling of the PV panels is stored in underground accumulation tanks, which makes it possible to increase the ground temperature around the tank to increase and restore the ground ability to provide heat after the heating season. Moreover, the PV panels will produce the electricity needed by the proposed system, as well as the electricity needed to satisfy the building heat demand. Due to the cooling of the PV panels, the efficiency of electricity production is expected to rise by 5–10%. The heat stored in the ground and in the storage tanks will be used by the heat pump to supply the building with heat. In addition, the application of the thermal energy storage system will raise the heat pump COP. Due to the use of advanced sun-tracking systems for PVT cells and solar collectors, the value of the average annual COP is expected to be higher than four. As a result, less electricity is required for the compressor operation in the heat pump cycle compared to typical ground source heat pumps. An additional advantage of the system is that, due to the use of accumulation tanks, the number of boreholes (and consequently—the ground heat exchanger surface area) is smaller than in traditional solutions. This will make it possible to reduce material costs of the lower heat source installation.

The system is fully based on RES because the electricity supplied to the heat pump will be generated by PV panels. Another novelty is the application of the sun-tracking system in the PV panels to increase the amount of electricity produced during the day. Moreover, the solar collectors equipped with a sun-tracking system are used to assist domestic hot water production in the spring–autumn period.

The proposed system makes it also possible to utilize the waste heat from the air conditioning system. The warm air downstream the condenser will be used to heat domestic hot water using a plate-and-tube heat exchanger with a developed surface. The heat can also be used to raise the temperature of the underground thermal energy storage and for domestic hot water preparation (this will be an expansion of the existing system). Consequently, an increase is expected in the system overall efficiency and in the ground source heat pump performance. In warm climates this modification can be used for domestic hot water heating, while in countries with moderate climates (Poland, the Czech Republic, Germany) the heat pump power output will be increased in the heating season. To make the system work as efficiently as possible, the ground should be insulated from above.

The proposed system is a comprehensive solution. The customer gets a ready-made building-heating system based entirely on RES. The system will be offered with the necessary systems of control and automation; it will be maintenance-free and require no action on the part of the end user. Furthermore, the system is autonomous as it generates the electricity needed to power the components. The demand for primary energy of a building equipped with the proposed energy system is zero, which means that the non-renewable primary energy demand (PED) factor [kWh/(m$^2$ year)] for the building energy supplies can be reduced to 0.

A prototype of the proposed system is installed in Limanowa (Poland), at the headquarters of the F.H.U Urządzenia Chłodnicze Czamara company. It supplies heat and electricity to the building. The system thermal and electric power capacity totals 75 kW and 41 kW, respectively. The RESHeat system will make it possible to carry out the following tests and analyses:

(a) Testing of thermal and electrical efficiency of solar photovoltaic cells combined with water-to-water heat pump systems

The RESHeat system makes use of intelligent sun-tracking PVT solar panels to improve the cooling efficiency of photovoltaic panels and increase net electricity production. The application of the sun-tracking mechanism in PVT panels enables a 30% rise in the production of electricity. An additional improvement in the efficiency of solar-to-electric energy conversion is achieved due to the cooling of PVT cells. The waste heat recovered from PVT cells can be used as the lower heat source of the heat pump. It can also be accumulated in the ground for use during the heating season.

(b) Analysis of the ground thermal regeneration due to PVT panels and utilization of waste heat from solar collectors

The novel idea is to use waste heat from solar collectors and PVT cells for ground regeneration (boreholes and the non-insulated heat accumulation tank). Solar collectors equipped with a sun-tracking system can heat domestic hot water up to 30 °C even in winter. This waste heat can also be used as the lower heat source of the heat pump.

(c) Analysis of the impact of the ground regeneration process on the operation of the ground source heat pump (changes in the COP in consecutive heating seasons)

The RESHeat system enables seasonal storage of thermal energy in the ground and uses the energy to improve the COP of the water-to-water heat pump by increasing the temperature of the lower heat source. Another very important thing is that ground regeneration due to the use of waste heat from the cooling of PV panels makes it possible to keep the heat pump COP constant from one heating season to the next. The heating system applied at the headquarters of the F.H.U. Urządzenia Chłodnicze Czamara company made use of a heat pump which had already been operating for five years without thermal energy storage in the ground. During that time, the heat pump coefficient of performance dropped from 3.1 to 2 because the ground was not regenerated sufficiently. Owing to the installation of the RESHeat system with ground regeneration, the COP increased from 2 to over 3.5 in 2020.

## 4.2 Components of the System

### *4.2.1 PVT Panels with a Sun-Tracking System*

The innovative photovoltaic cooling was developed within the HySOL project *High-efficiency hybrid solar system for generation of thermal and electric energy used in buildings*, financed from the means of the Polish-German Sustainability Research (STAIR) Foundation. The HySOL project started in October 2017 and is to be completed by the end of March 2021. The main goal of the HySOL project is to develop a novel PV cooling system that is easy to install and can be applied in solar systems. To test the proposed solution, a fully operational prototype was built at the Faculty of Mechanical Engineering of the Cracow University of Technology (cf. Figs. 4.3 and 4.4.

Figure 4.4 shows a fragment of the roof surface layout with marked location of PV panels, both stationary and those equipped with the sun-tracking mechanism. The stationary panels face directly south and are inclined at 29° angle to the ground surface taking account of the roof slope. The following are installed in the room under the PV panels: heat pumps for the panels cooling, an inverter together with a

Fig. 4.3 Location of the test installation for cooling PV panels with a sun-tracking system

Fig. 4.4 Location of stationary and sun-tracking PV modules on the roof of the building

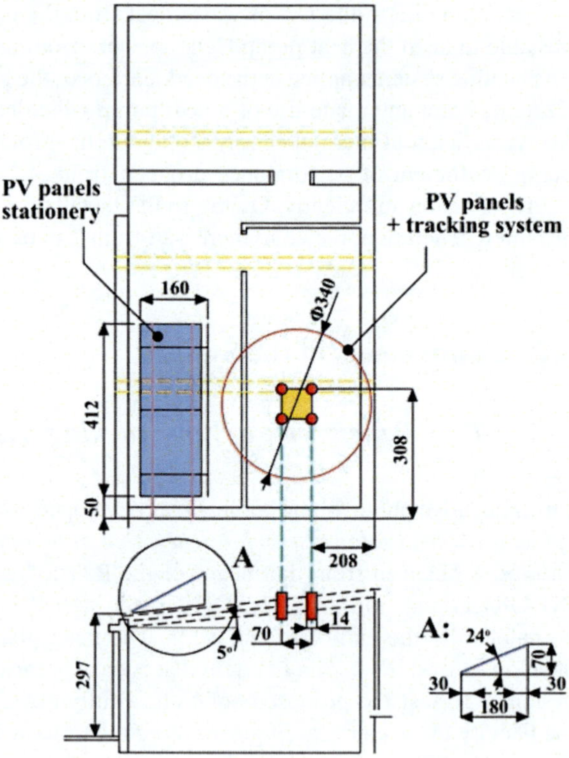

**Table 4.1** Eco Delta ECO-280P panels—specification for standard test conditions (STC)

| Type | Eco Delta ECO-280P panels |
|---|---|
| Exterior dimensions: | 1,640 mm × 992 mm × 35 mm |
| Maximum power | 280 W |
| Maximum power voltage | 33.30 V |
| Maximum power current | 8.41 A |
| Open-circuit voltage | 38.60 V |
| Short-circuit current | 9.00 A |
| Module efficiency | 17.21% |
| Maximum voltage | 1,000 V |
| Maximum rated voltage of the series fuse | 15 A |

battery bank and a supervisory control and data acquisition (SCADA) system (Table 4.1).

The test stand shown in Fig. 4.5 is composed of eight polycrystalline PV panels (1), four of which are installed on the sun-tracking system (2) and the other four are stationary. The sun-tracking system used on the test stand is an improved design developed by the Polish manufacturer ELFRAN, a consortium member of the HySOL project.

The PV panel cooling system adapted to the test stand consists of radiators mounted directly on the back of the PV panels and two heat pumps (3) with an advanced control system supplying the coolant—a 51% aqueous solution of propylene glycol—to the storage tank (5). The heat pumps are installed only to ensure a constant operating condition of the cooling system—a constant inlet temperature. The implemented control system regulates the circulation pumps (7) within the cooling cycles. The pumps are activated when the surface temperature of the PV panels exceeds 25 °C. On the one hand, this control strategy makes it possible to reduce the heat loss from the system when the PV panels reach temperatures well below 25 °C. In this case, cooling is not required; it may even reduce the efficiency of the PV panels. On the other hand, the adopted control strategy enables the system operation at a temperature optimal for the PV panels.

Figure 4.5a presents the prototype of the cooling radiators mounted on photovoltaic panels It should be emphasized that the presented solution is a unique combination of simultaneous solar heating and PV cooling in one device. For this reason, the experimental testing and numerical simulation of such a unique structure is the main novelty of the concept. Furthermore, the radiator size is adjustable and can be adapted to different dimensions of PV panels offered by different manufacturers. Moreover, the proposed cooling system design is absolutely not in collision with the guarantee terms of PV panels. It can therefore be used in photovoltaic systems on a large scale. A detailed diagram of the test system is presented in Fig. 4.6.

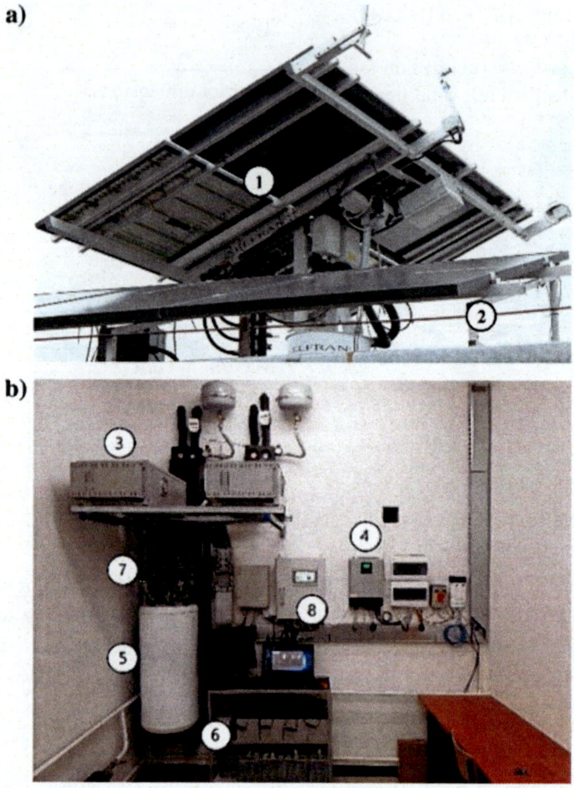


**Fig. 4.5** Test stand to test the cooling system of PV panels equipped with the sun-tracking system: **a** PV panels mounted on the roof

The results of efficiency measurements of the PVT cells are presented in Figs. 4.7 and 4.8. The measurements were performed in October 2018. Based on them, it can be concluded that PV cooling increases electricity production by about 4% to 6% on sunny days. The rise in electricity production is even more significant if the sun-tracking system is applied. The increase is due to higher efficiency of the solar energy utilization, as the sun-tracking system sets the PV panel in a position perpendicular to the incident solar radiation. In this way, more solar energy is available to PV panels with the sun-tracking system compared to stationary panels.

### 4.2.2 *Demonstration Installation of a Zero-Emission System with Heat Accumulation in the Ground*

The developed PVT panels are an important part of a large RES solution installed in Limanowa (Poland). The Limanowa system has been operating since August 2019. It is visualized in Fig. 4.9, and the system diagram is presented in Fig. 4.10. A photo of the facility is shown in Fig. 4.11. The system was created under the *SOPSAR* project

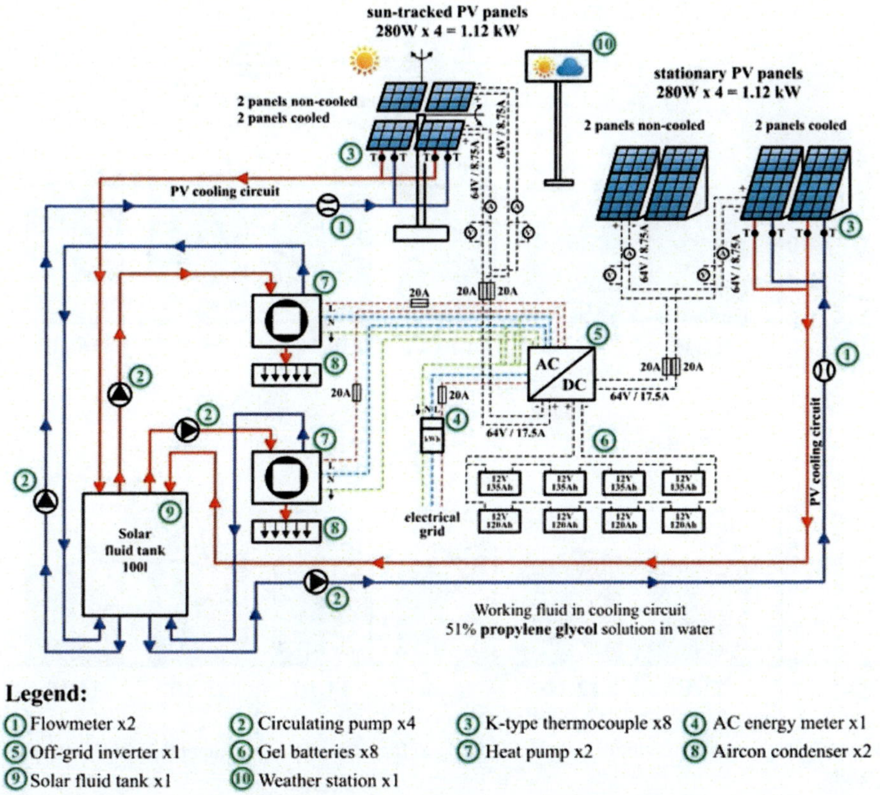


**Fig. 4.6** Diagram of the test stand with marked location of measuring points

performed within the NCRD program *A fast track for small and medium enterprises*. The project main beneficiary is the F.H.U Urządzenia Chłodnicze Czamara company. The Department of Energy of the Cracow University of Technology assisted the company in designing the system.

The installation in Limanowa is made of the following components:

**A**—solar collectors with a sun-tracking system (10 kW thermal power) supplying the domestic hot water tank (central heating),

**B**—solar collectors with a sun-tracking system (15 kW thermal power) supplying the domestic hot water tank (central heating),

**C**—PVT cells (made by the ELFRAN company) with a sun-tracking system with the total electric power of 16 kW,

**D**—102 stationary PVT panels (made by the ELFRAN company) with the total electric power of 25 kW,

**E, F**—heating water buffers (central heating) with the capacity of 1,000 L (E) and 700 L (F),

**G**—a domestic hot water tank (central heating) with the capacity of 800 L,

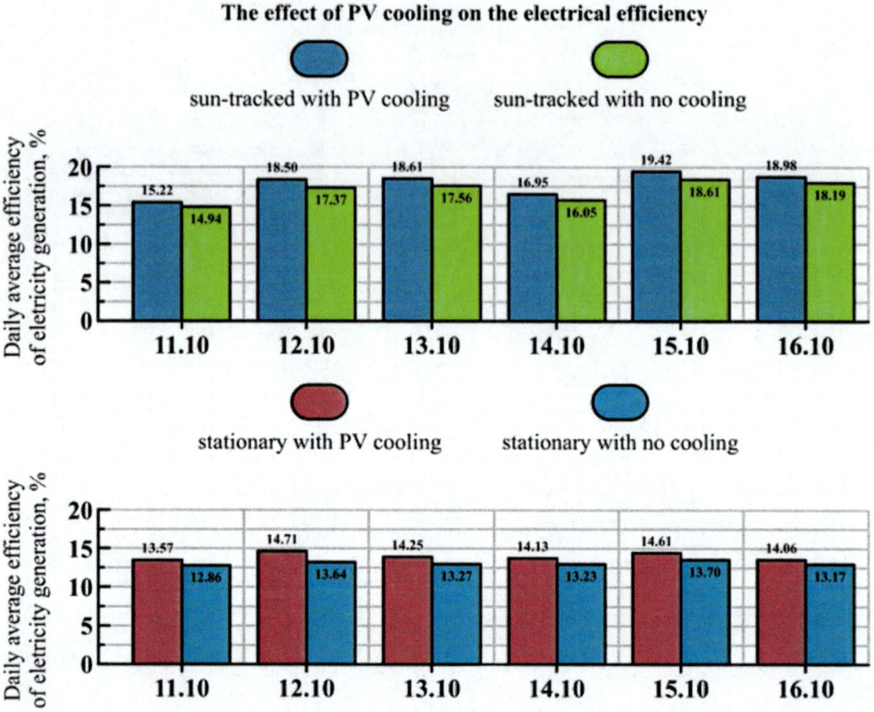


**Fig. 4.7** Impact of PV cooling on the daily average efficiency of sun-tracking panels and stationary panels

**H**—a ground source heat pump with the heating power of 50 kW,
**I**—an underground, thermally insulated storage tank (V = 35 $m^3$) to store thermal energy from the solar collectors,
**J**—an underground, thermally insulated storage tank (V = 35 $m^3$) to store thermal energy from the PVT panels,
**K**—ground heat exchangers with developed (finned) surfaces,
**L**—an indirect heat exchanger for thermal energy storage.

Additionally, the control system and the SCADA system are adapted to monitor the performance of the installation and to select the best heat source for the heat pump. The solar installation in Limanowa, shown in Fig. 4.10, covers the demand for heat (underfloor heating), cool (underfloor cooling) and electricity of an 860 $m^3$ office building and a 1.225 $m^3$ factory. A ground source water-to-water heat pump is responsible for the building heating, which is done using a system of underfloor heating. The demand for heat, cool and electricity is fully covered by the installation.

The installation is served by a ground source water-to-water heat pump. The accumulation tanks were installed in March 2019. However, the entire installation became fully operational in November 2019. Due to some delays, it was not possible to complete the installation in May, which is the best month for the system operation

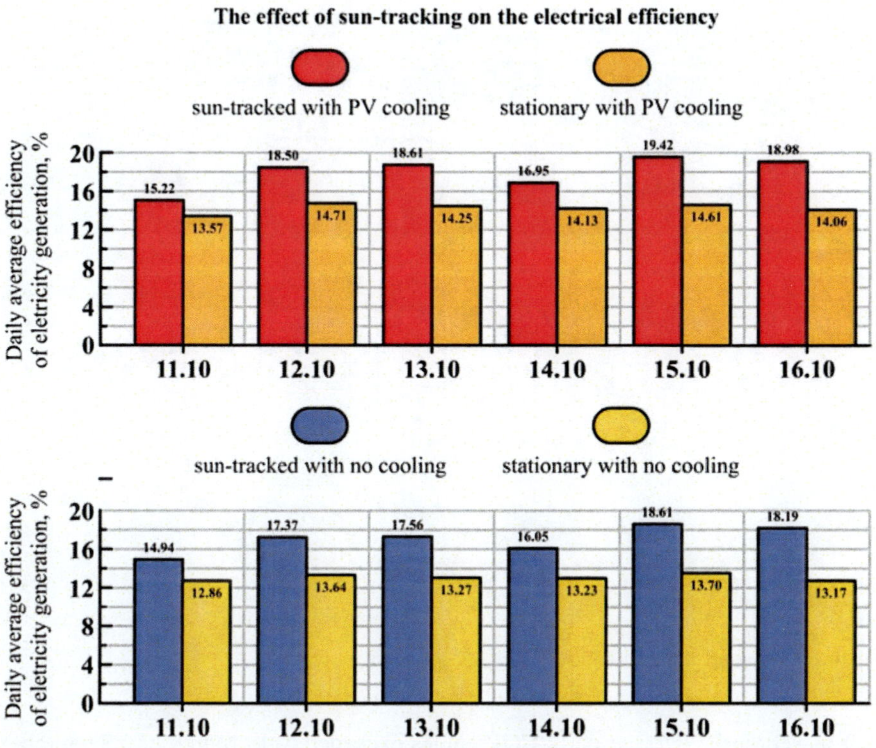


**Fig. 4.8** Impact of sun-tracking on the daily average efficiency of panels both with and without cooling

**Fig. 4.9** Visualization of the RESHeat system installed at the headquarters of the FHU Urządzenia Chłodnicze Czamara company

to begin because it is then possible to fully utilize the waste heat from the PVT cells and solar collectors accumulated during the spring–autumn period. The values of the heat pump average daily COP coefficients for selected days are presented in the charts below for the following months: February (Fig. 4.12), March (Fig. 4.13, September (Fig. 4.14), October (Fig. 4.15), November (Fig. 4.16). It is worth noting

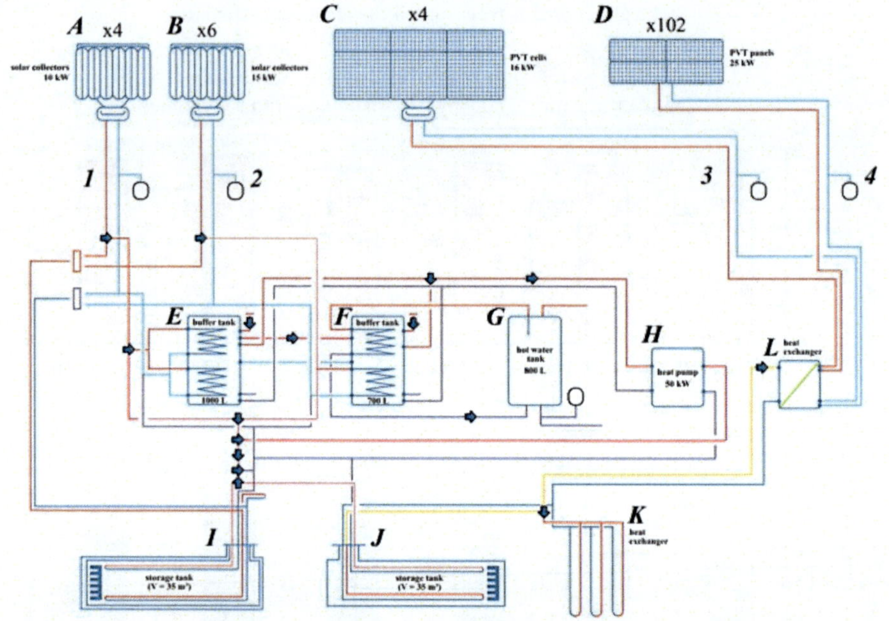


**Fig. 4.10** Diagram of the RESHeat system in Limanowa

that in February 2020 the daily COP values exceeded 4. In March and September 2020 they were higher and reached even 4.5. In October and November 2020 the average daily COP values were below 4. The analysed period is the first year of the system operation. The ground is not sufficiently heated yet, so the monthly average COP values are lower than the target values.

Figure 4.17 shows the amount of heat $Q$ in kWh produced by the heat pump and the amount of electricity $E_l$ consumed by the compressor for the periods from February to April and from September to December. The measurement data acquisition started in February 2020. The measured quantities were the amount of electricity consumed by the heat pump and the thermal energy it produced. The heat pump consumed the most thermal energy in the November–December period. This is due to the FHU Urządzenia Chłodnicze Czamara company expansion in the summer, which created additional space to be heated. Figure 4.18 presents the values of the COP averaged over consecutive months of the year. The highest value of the average monthly COP was obtained for April: COP = 3.8 and the lowest for December: COP = 2.9.

Figure 4.17 shows the amount of heat $Q$ in kWh produced by the system over consecutive months in 2020. In addition, the amount of electricity (kWh) consumed by the heat pump compressor is presented. Figure 4.18 shows the heat pump average monthly coefficient of performance, which totalled 3.46 for February, 3.6 for March, 3.83 for April, 3.5 for September, 3.38 for October, 3.04 for November and 3.00 for December. After the ground full regeneration in the next year, higher average monthly values of the COP are expected.

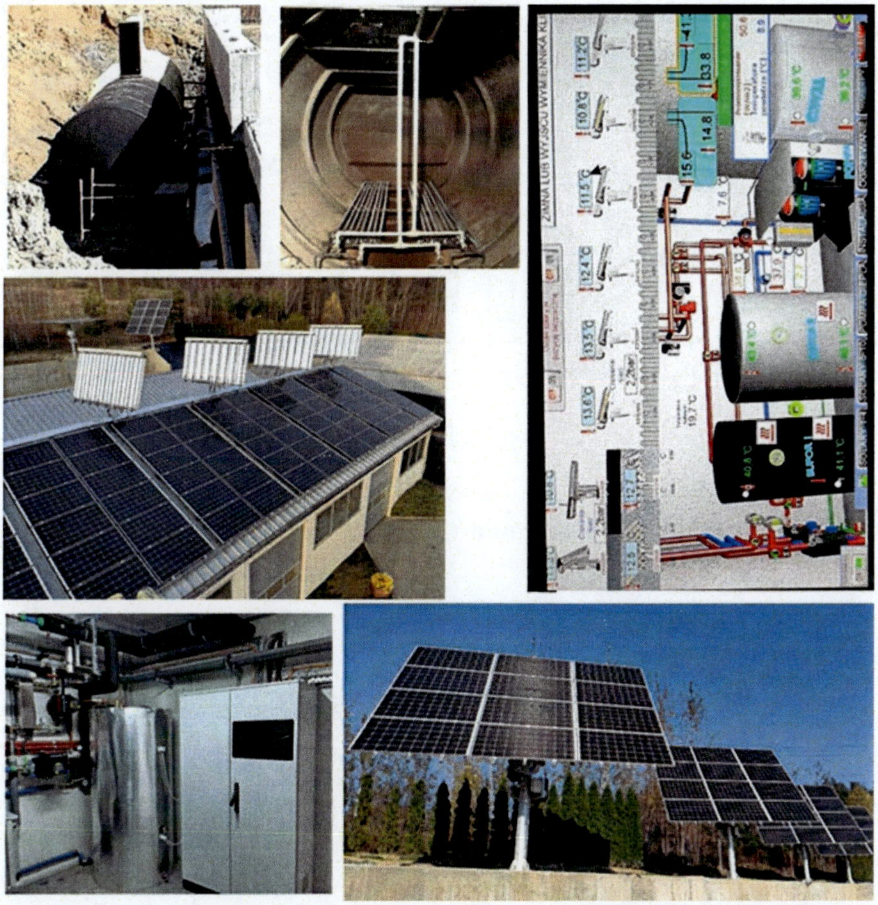

**Fig. 4.11** RESHeat system installed at the headquarters of the FHU Urządzenia Chłodnicze Czamara company in Limanowa

Figure 4.19 shows the thermal energy values obtained in the period from May to December from the cooling of the stationary PVT cells, the PVT cells with the sun-tracking system and from the rotating collectors.

Analysing Fig. 4.19, it can be seen that by far the largest part of thermal energy comes from the rotating solar collectors (ELFRAN Revolution Solar 8). The PVT cells equipped with the sun-tracking system are also efficient. But the thermal efficiency of the stationary PVT cells is very low.

Figure 4.20 presents the amount of electricity (in kWh) produced by the 16 kW PVT panels with the sun-tracking system and the stationary 25 kW PVT panels. Under weak insolation, the PVT panels equipped with the sun-tracking system produce much more electricity even though their power capacity is lower compared to the stationary PVT panels.

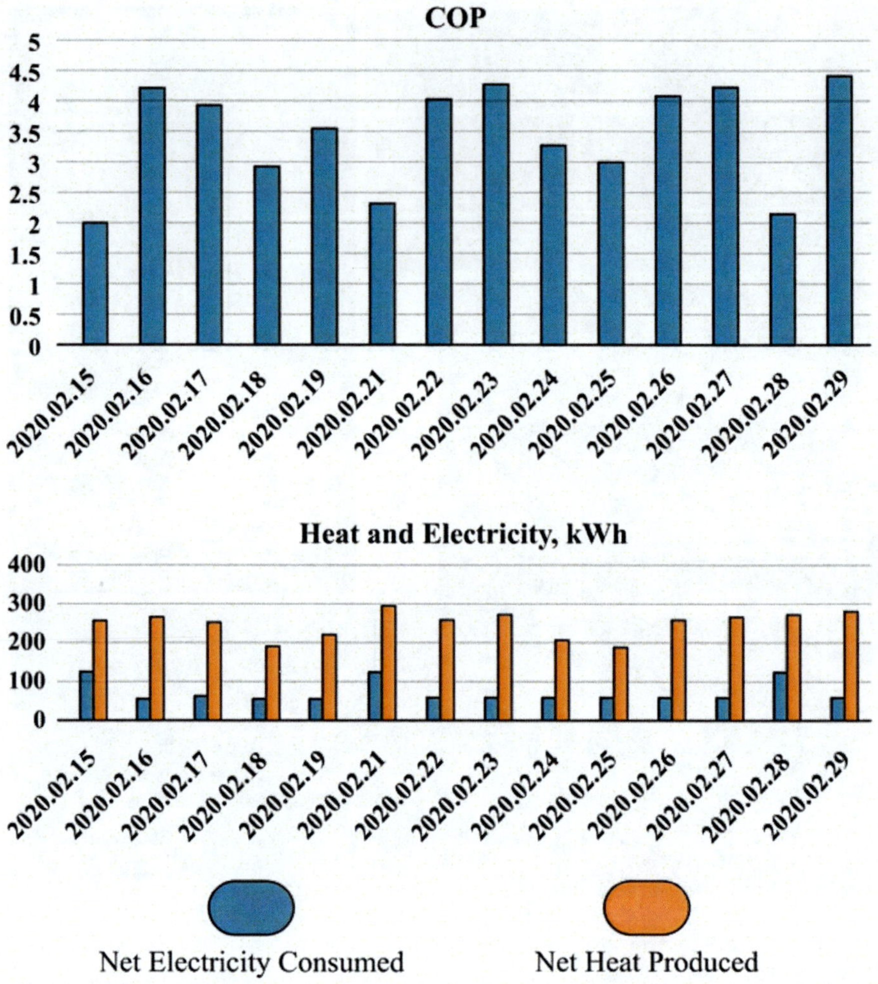


**Fig. 4.12** Average daily COP of the heat pump and the thermal energy (kWh) produced by the heat pump and the electricity (kWh) consumed by the heat pump compressor on selected days in February 2020

The results of the measurement data analysis point to significant reserves in the system operation. The 2021 measurement campaign will enable full determination of how the ground regeneration due to utilization of waste heat from the PVT cells and solar collectors will affect the value of the annual average COP. This will be possible once the system has operated through an entire spring–autumn period with waste heat delivered from the solar installation to the storage tanks and to the ground.

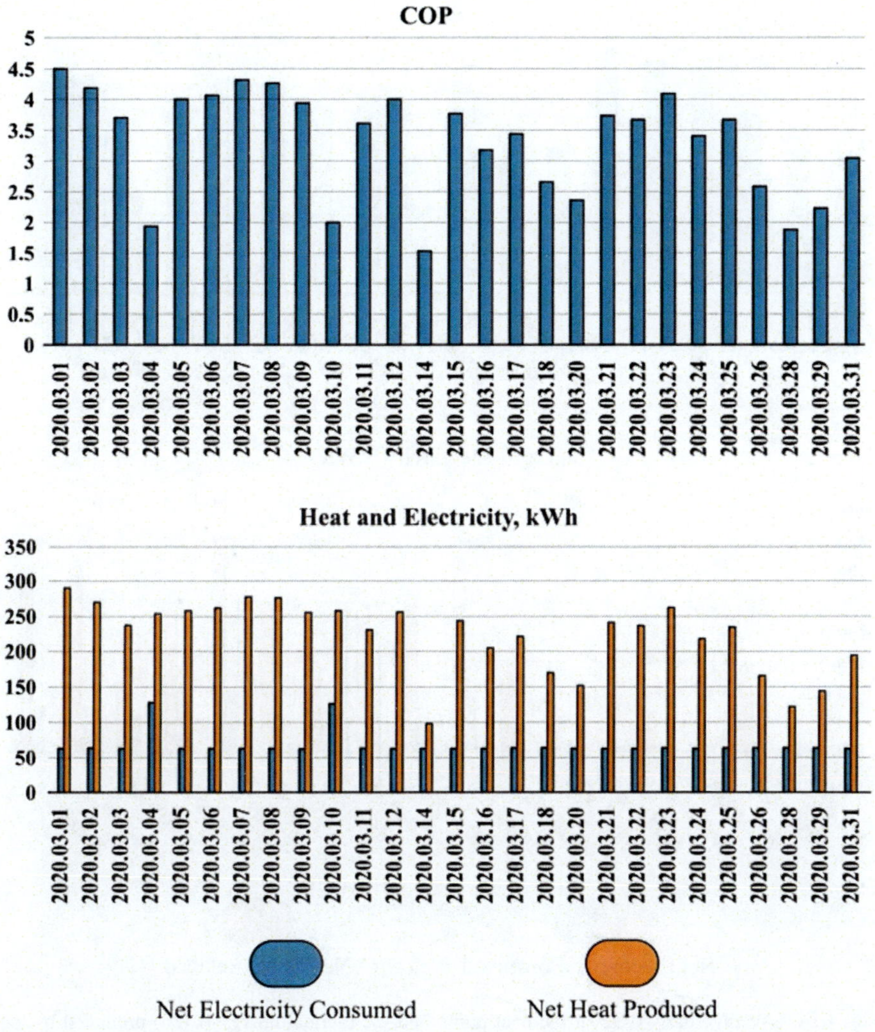


**Fig. 4.13** Average daily COP of the heat pump and the thermal energy (kWh) produced by the heat pump and the heat and electricity (kWh) consumed by the heat pump compressor on selected days in March 2020

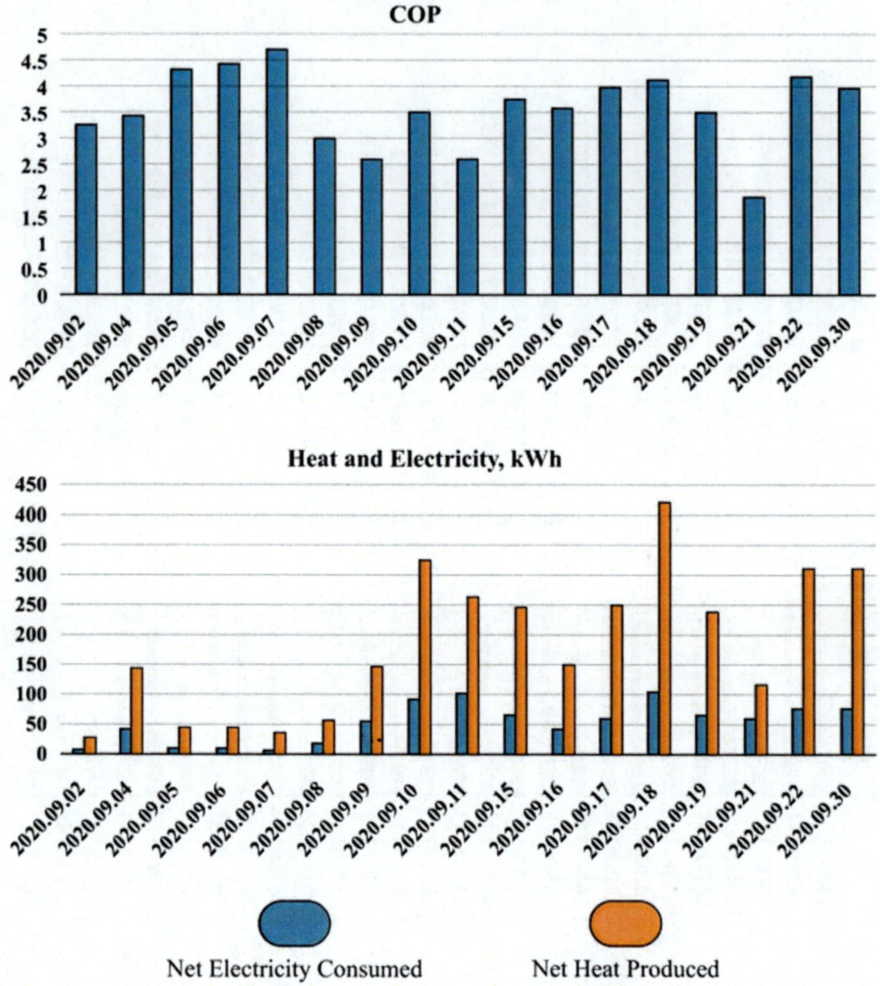


**Fig. 4.14** Average daily COP of the heat pump and the thermal energy (kWh) produced by the heat pump and the electricity (kWh) consumed by the heat pump compressor on selected days in September 2020

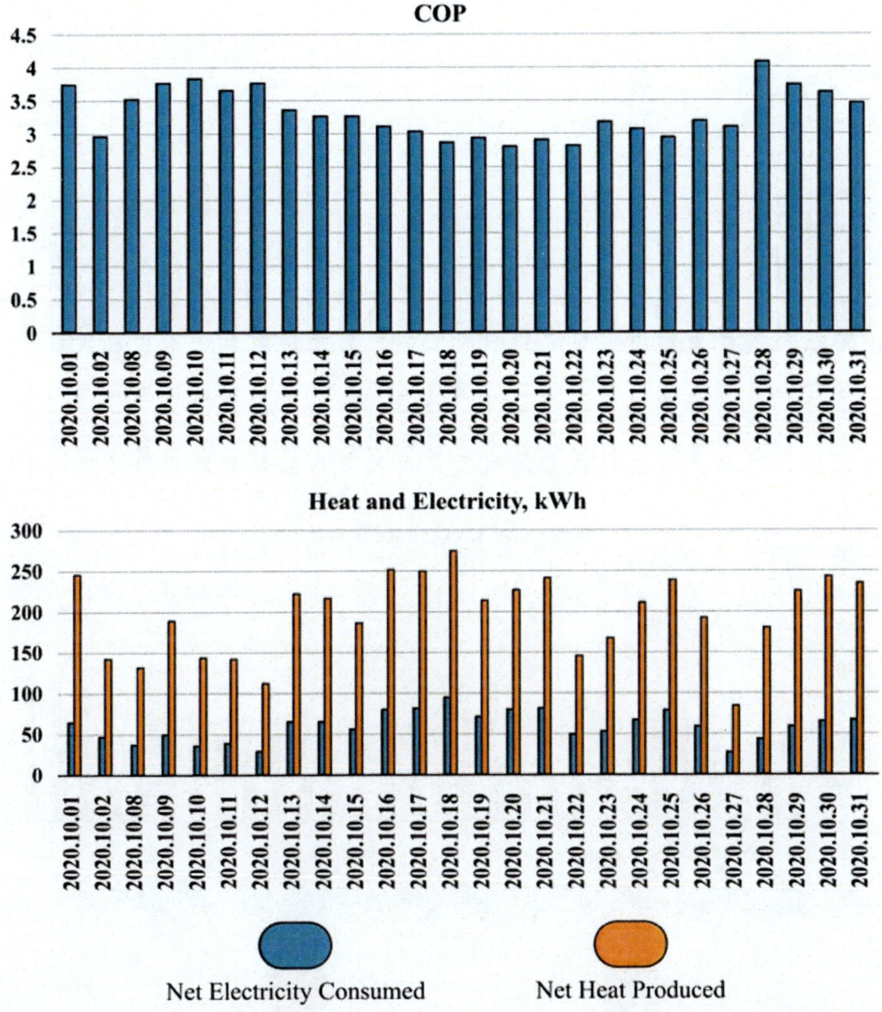


**Fig. 4.15** Average daily COP of the heat pump and the thermal energy (kWh) produced by the heat pump and the heat and electricity (kWh) consumed by the heat pump compressor on selected days in October 2020

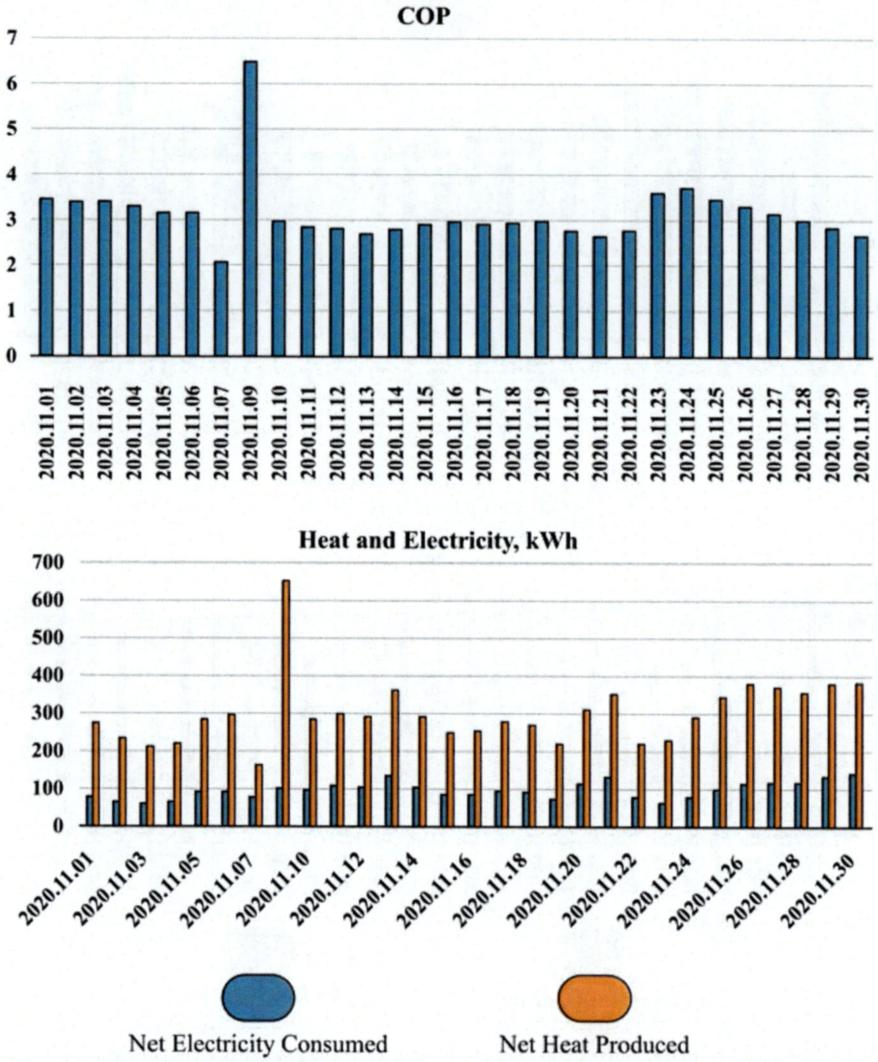


**Fig. 4.16** Average daily COP of the heat pump and the thermal energy (kWh) produced by the heat pump and the heat and electricity (kWh) consumed by the heat pump compressor on selected days in November 2020

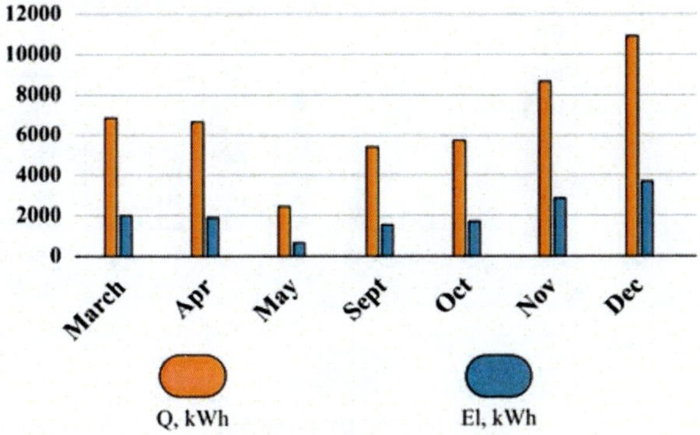


**Fig. 4.17** Thermal energy $Q$ (kWh) produced by the heat pump and electricity (kWh) consumed by the compressor

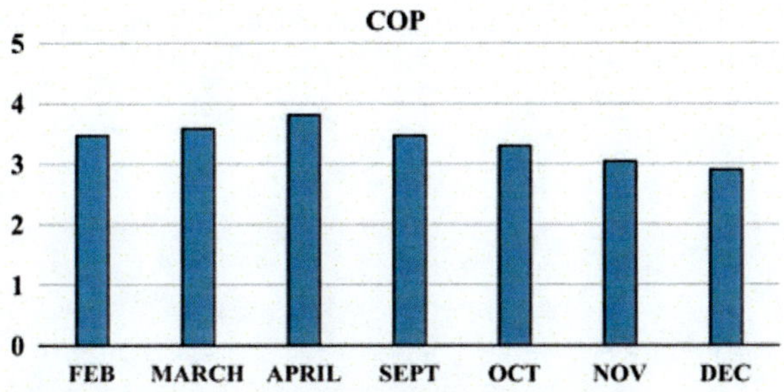


**Fig. 4.18** Average monthly COP of the heat pump

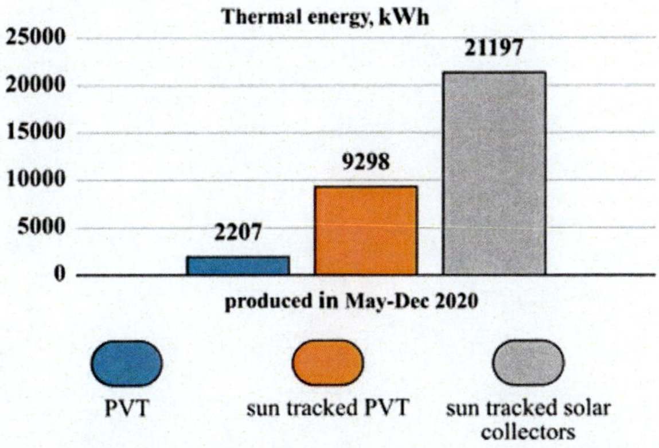


**Fig. 4.19** Thermal energy produced from the cooling of the PVT cells (stationary and with the sun-tracking system) and from the solar collectors from May to December 2020

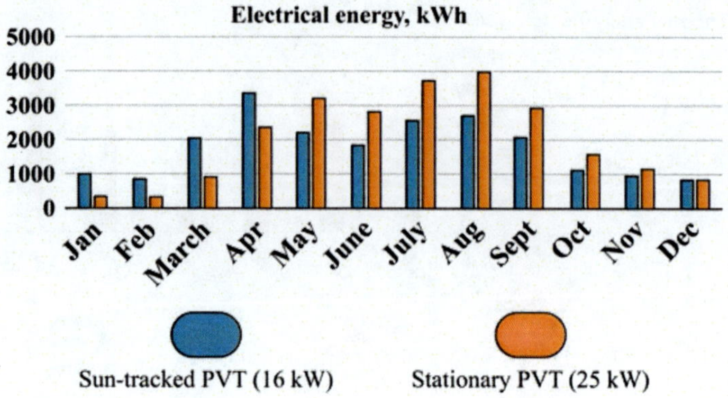


**Fig. 4.20** Amount of electricity produced by PVT modules with the sun-tracking system and with stationary PVT panels

# Chapter 5
# Mathematical Modelling of the Resheat System

In the proposed system, effective regeneration of the ground is achieved by storing waste heat from PVT panels in a non-insulated storage tank and/or boreholes. This enables heat transfer from the storage tank/borehole to the ground and an increase in the ground temperature in the summer season. Since the ground is insulated from above, heat losses are negligible and the ground temperature rises. Due to the heat transfer from the storage tank or boreholes to the ground, the ground temperature in winter is higher compared to the situation when waste heat from PVT cells is not recovered. The heat will be stored in the ground, increasing the possibility of transferring heat to the heat pump during the winter season.

The control system of the RESHeat solution works as follows: the system measures the temperature of the ground heat source (non-insulated tank, insulated tank or boreholes) and uses heat from the source with the highest temperature as the heat pump heat source. If this temperature is higher than the heat pump allowable value, the flows of heat from different heat sources are mixed using mixing valves. This approach makes it possible to achieve the highest possible energy efficiency of the heat pump.

The analysis covers a building heating system based on a heat pump, solar collectors, rotating PVT panels and an accumulation tank. The waste heat from the cooling of PV panels is stored in the ground during the spring–summer period. A mixture of water and glycol is used as an intermediate medium to cool the PV panels. The waste heat is then transferred to the heat pump evaporator.

The soil essential property affecting the heat transfer process in the ground is its thermal conductivity. This quantity is very important for designers as the main determinant of the intensity of the heat transfer between the accumulation tank and the soil. Thermal conductivity can be calculated by an algorithm that takes account of the influence of various factors (bulk density, clay content, temperature, humidity). Owing to such an approach, it is possible to take the variability of this parameter in consideration, in relation to others. In the development of such an algorithm, usually



P. Ocłoń, *Renewable Energy Utilization Using Underground Energy Systems*, Lecture Notes in Energy 84, https://doi.org/10.1007/978-3-030-75228-6_5

only the most important factors that are taken into account, but the more of them are included in the equation, the more accurate the results will be.

## 5.1 Thermal Properties of the Ground

The parameters that should be taken into consideration in the first place are the physical and geometrical properties (Rerak 2020), as they have an essential effect on heat conduction. The most significant quantity is the moisture content in the soil and the soil density in the dry state. Obviously, these properties involve changes in other characteristics because the quantities describing soil are often coupled with each other. Humidity is the parameter describing the soil water content. The soil density in the dry state determines the soil weight at a given volume—the dry state is achieved by drying in the temperature of 105 °C, whereas volume is determined in the normal state.

The soil density in the dry state and the soil moisture content have a significant impact on thermal conductivity. The higher their values, the higher the soil heat transfer coefficient. Moreover, the heat transfer coefficient of the soil is affected by parameters such as temperature or the soil mineral composition.

The rise in thermal conductivity as a function of the moisture content can be explained easily. By getting between individual particles, water fills the intermolecular spaces of the soil. The effective contact area is increased and the heat transfer is improved. Once the soil maximum water saturation is achieved, the increase in the soil thermal conductivity is stopped (Rerak 2020).

The increase in the soil thermal conductivity as a function of the moisture content is shown in Fig. 5.1. The effect of the soil content on its thermal conductivity depends on the mineral composition. These relationships can be determined empirically. For

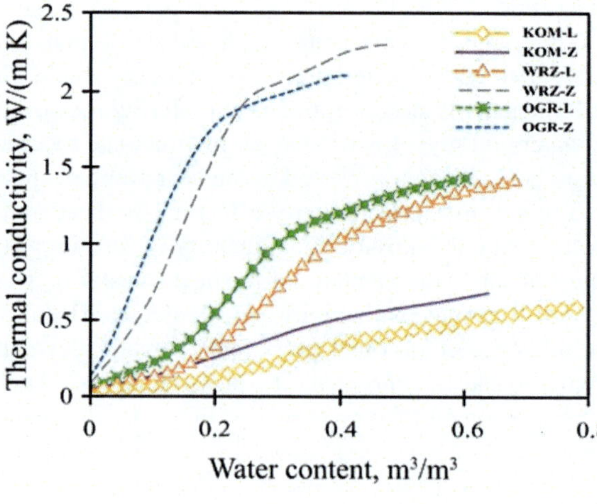


**Fig. 5.1** Chart illustrating the soil thermal conductivity depending on its moisture content (Usowicz, 2015)

**Fig. 5.2** Chart illustrating the soil thermal conductivity for different density values depending on the moisture content (Usowicz, 2015)

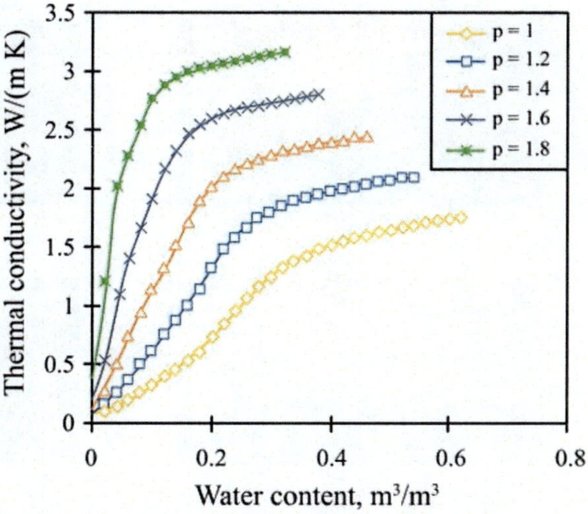


different soils, the dependence of the heat conductivity coefficient on the moisture content takes a different form (Usowicz, 2015).

Figure 5.1 presents the dependence of the soil thermal conductivity on its moisture content for the following kinds of soil (Usowicz, 2015):

- compost soil: loose (marked as KOM-L) and compacted (marked as KOM-Z),
- heather soil: loose (marked as WRZ-L) and compacted (marked as WRZ-Z),
- garden soil: loose (marked as OGR-L) and compacted (marked as OGR-Z).

As the dry-state soil density increases, the number of particles per unit volume increases as well. The rise in the number of particles per unit volume is also associated with an increase in the number of so-called contact points per unit volume. This increases the contact area, which facilitates the heat transfer. It can be seen in Fig. 5.2 that compacted soil has higher thermal conductivity compared to the same soil type in the loose state. A twofold difference in the change in the soil thermal conductivity is observed between loose and compacted types for the heather soil with the moisture content of 40% (Rerak 2020).

Figure 5.3 presents the change in the soil thermal conductivity depending on the moisture content for different values of the soil density. The analysed soil is sand made in 100% from quartz. An increase in the sand density by 80% ($\rho = 1$ kg/m$^3$ versus $\rho = 1.8$ kg/m$^3$) resulted in a rise in the soil thermal conductivity by over 100%. It is worth noting that there is no single relationship that is valid for all soil types (Rerak 2020).

An important factor with an impact on the soil thermal conductivity is the soil mineral structure. As the composition of the soil changes, the soil thermal conductivity changes as well. From one place to another, it may be higher or lower. It is therefore impossible to precisely determine the soil thermal conductivity in the entire cross-section because it is impossible to establish the shares of all its components

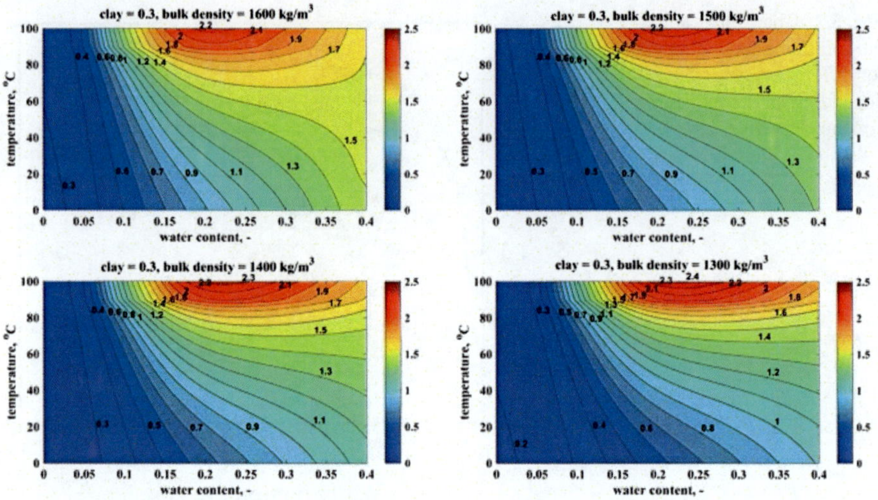


**Fig. 5.3** Values of the ground heat conductivity coefficient calculated based on the Campbell-de Vries model for the clay mass fraction of 0.3

in every point quantitatively. To include this factor in the calculation, an indirect method can be used (Rerak, 2020). Parameters such as the shape and size of particles as well as their arrangement are taken into account. However, there are advanced mathematical models, such as the Campbell model for example (Campbell et al., 1994), intended for determination of the soil thermal conductivity as a function of the following parameters: temperature, moisture content, bulk density and clay content.

### *5.1.1 Thermal Conductivity of the Ground—the Campbell-De Vries Model*

An advanced model for calculating the ground thermal conductivity was published by Campbell et al. (1994); it is a modification of the original de Vries model (de Vries, 1963). The model is based on the assumption that the material thermal conductivity is determined by the weighted sum of thermal conductivities of the material components:

$$k = \frac{w_w \theta k_w + w_s \phi_s k_s + w_g \phi_g k_g}{w_w \theta + w_a \phi_s + w_m \phi_g} \tag{5.1}$$

where $w_g$, $w_s$ and $w_w$ are weighting factors for gas, solid and water; $k$ is the heat conductivity coefficient, and $k_g$, $k_s$ and $k_w$ are heat conductivity coefficients for gas, solid and water, respectively. The volume fractions of gas, solid and water are $\phi_g$,

$\phi_s$ and $\theta$, respectively. The gaseous phase thermal conductivity corresponds to both the air conductivity ($k_a$) and the component of latent heat released due to evaporation and condensation. In most soil models, thermal conductivity is considered to be independent of temperature. However, Campbell et al. (1994) proved that $k$ is temperature-dependent, and this dependence is mainly due to the changeability of the latent heat transport depending on temperature. Compared to the de Vries model (de Vries, 1963), the model proposed by Campbell et al. (1994) offers a more reasonable treatment of the transition from the high to the low content of water vapour as the soil drying proceeds. The de Vries model (de Vries, 1963) takes account of thermal conductivity of continuous phases: when the soil is saturated, the continuous phase is liquid water; when the soil is dry, the continuous phase is gas. However, the de Vries model (de Vries, 1963) did not take account of intermediate conditions of the soil partial saturation. This was introduced by Campbell et al. (1994) by defining the continuous function using a variable referred to as "fluid" thermal conductivity, which can be used over the entire range of the water content. The authors defined the "fluid" thermal conductivity variable as:

$$k_f = k_g + f_w(k_w - k_g) \tag{5.2}$$

where $f_w$ is an empirical parameter defined as:

$$f_w = \frac{1}{1 + \left(\frac{\theta}{\theta_o}\right)^{-q}} \tag{5.3}$$

ranging from 0 in dry soil to 1 in saturated soil. Parameters $\theta_0$ and $q$ are the material-specific properties that define when the water content starts to affect thermal conductivity and the rate of transition from air-dominated conductivity to conductivity dominated by water. It is found that parameters $q$ and $\theta_o$ are strongly correlated with the clay content, and the following regression equations are presented:

$$q = 7.25m_y + 2.52, \tag{5.4}$$

$$q = 0.33m_y + 0.078, \tag{5.5}$$

where $m_y$ is the clay fraction content [0–1].

The next step is to calculate relevant weighting factors:

$$w_g = \frac{1}{3}\left[\frac{2}{1 + g_a\left(\frac{k_g}{k_f} - 1\right)} + \frac{1}{1 + g_c\left(\frac{k_g}{k_f} - 1\right)}\right], \tag{5.6}$$

$$w_s = \frac{1}{3}\left[\frac{2}{1 + g_a\left(\frac{k_s}{k_f} - 1\right)} + \frac{1}{1 + g_c\left(\frac{k_s}{k_f} - 1\right)}\right], \tag{5.7}$$

$$w_w = \frac{1}{3}\left[\frac{2}{1 + g_a\left(\frac{k_w}{k_f} - 1\right)} + \frac{1}{1 + g_c\left(\frac{k_w}{k_f} - 1\right)}\right], \tag{5.8}$$

where $g_{\mathrm{a}}$ and $g_{\mathrm{c}}$ are the shape factors. The value of $g_{\mathrm{a}}$ for mineral soil is 0.088, and $g_{\mathrm{c}} = 1{-}2g_{\mathrm{a}}$. Therefore only one shape factor has to be obtained.

The gaseous phase thermal conductivity $k_{\mathrm{g}}$ is expressed by the following equation:

$$k_g = k_a + \frac{L_\upsilon \Delta f_w \hat{\rho} D_\upsilon}{P - e_a}, \tag{5.9}$$

where $k_{\mathrm{a}}$ is the thermal conductivity of air, $L_\upsilon$ is the latent heat of evaporation, $\Delta f_w$ is the slope of the saturated vapour pressure function, $D_\upsilon$ is the water vapour diffusivity for soil, $\hat{\rho}$ is the molar density of air, $P$ is atmospheric pressure and $e_a$ is the real pressure of vapour. The second term denotes latent heat, which is responsible almost entirely for the dependence of thermal conductivity to soil on temperature. Figures 5.3, 5.4, 5.5, 5.6, 5.7 present values of the ground heat conductivity coefficient determined from the Campbell-de Vries model for the moisture content of 0–0.4, temperature from 0 to 100 °C, the clay content from 0.3 to 0.5 and bulk density from 1,300 kg/m$^3$ to 1,600 kg/m$^3$.

Analysing the values of the ground heat conductivity coefficient, it should be noted that the coefficient rises with a rise in bulk density, temperature and the content of moisture.

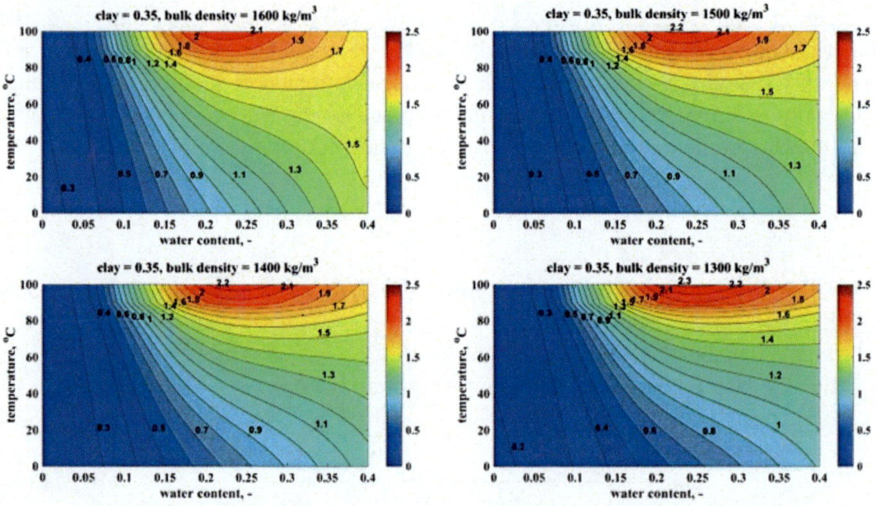


**Fig. 5.4** Values of the ground heat conductivity coefficient calculated based on the Campbell-de Vries model for the clay mass fraction of 0.35

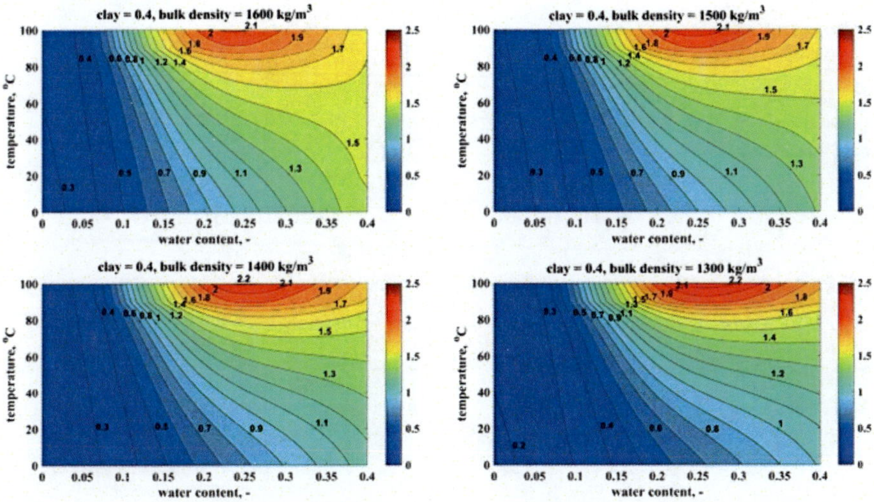


**Fig. 5.5** Values of the ground heat conductivity coefficient calculated based on the Campbell-de Vries model for the clay mass fraction of 0.4

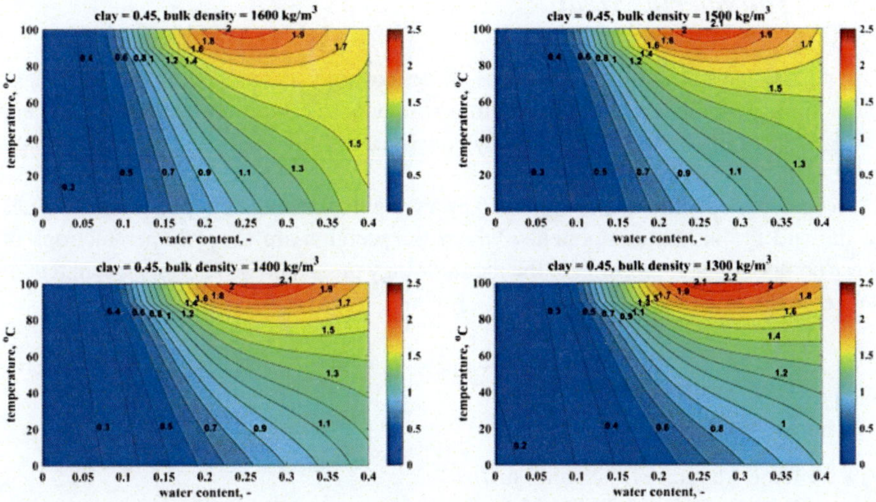


**Fig. 5.6** Values of the ground heat conductivity coefficient calculated based on the Campbell-de Vries model for the clay mass fraction of 0.45

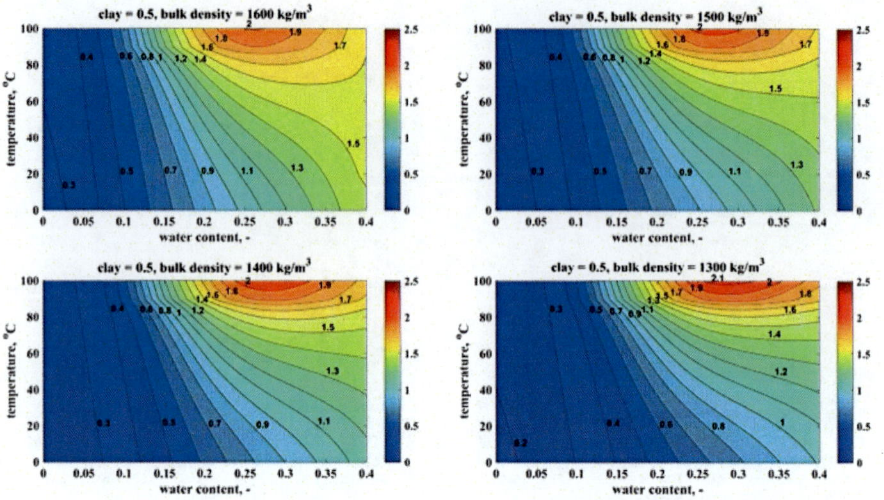


**Fig. 5.7** Values of the ground heat conductivity coefficient calculated based on the Campbell-de Vries model for the clay mass fraction of 0.5

### 5.1.2 Soil Specific Heat

Due to its internal structure, soil has to be considered as a porous, multi-phase medium. In general, it consists of a porous solid whose pores may be filled with water or air. Depending on the case, soil may consist of a solid, water and air (three-phase soil).

Volumetric specific heat is obtained as a weighted average of specific heat values of the soil individual components, where the weights are the volume fractions of these components. According to the methodology presented in (Campbell et al. 1994; de Vries 1963):

$$c_h = c_s\theta_s + c_l\theta + c_g\theta_g + c_o\theta_o \tag{5.10}$$

where

$\theta_s$ volume fraction of the solid, m³/m³
$\theta$ volume fraction of the liquid (volume moisture content), m³/m³
$\theta_g$ volume fraction of the gas, m³/m³
$\theta_o$ volume fraction of the organic matter, m³/m³.

while $c_s$, $c_l$, $c_g$, and $c_o$ are the values of the volume specific heat for the solid, liquid, gas and organic matter, which are read from Table 5.1.

Due to the small values of density and volumetric specific heat, the $C_g\theta_g$ term for the gas fraction is usually omitted. Very often the above formula is reduced even further, to the following form:

**Table. 5.1** Density, specific heat and thermal capacity at a constant volume for different components of the soil (Kodešová et al. (2013))

| Soil component | Density, t/m$^3$ | Specific heat at constant pressure, kJ/(kg K) | Volumetric thermal capacity, MJ/m$^3$/K |
|---|---|---|---|
| Minerals | 2.65 | 0.73 | 1.9 |
| Organic constituents | 1.3 | 1.9 | 2.5 |
| Water | 1 | 4.18 | 4.18 |
| Ice (0 °C) | 0.92 | 2 | 1.9 |
| Air | 0.0012 | 1 | 0.0012 |

$$C_h = C_s(1-\varphi) + C_l\theta \tag{5.11}$$

where $\varphi$ is the ground total porosity. The specific heat measuring data for sand are listed in Table 5.2.

## 5.2 Modelling the Building Heating System with Heat Accumulation in the Ground

### *5.2.1 Calculation of the Heat Pump Coefficient of Performance*

The energy efficiency ratio (EER) and the coefficient of performance (COP) define the heat pump heating and cooling power capacity. The higher their values, the more (heating or cooling) energy is delivered by the heat pump at a given energy consumption level. The EER defines the refrigeration capacity, whereas the COP—the heating capacity. For instance, the EER of 3.4 means that per one unit of consumed energy the device delivers 3.4 units of cooling energy. The EER and COP are found using the following formulae:

$$EER = \frac{Q_O}{L_S} \tag{5.12}$$

$$COP = \frac{Q_K}{L_S} = \frac{Q_O + L_S}{L_S} \rightarrow COP = EER + 1 \tag{5.13}$$

where

$Q_o$ heat absorbed in the evaporator
$Q_K$ heat given up in the condenser
$L_S$ energy consumed by the compressor.

**Table. 5.2** Sand specific heat for temperatures from −18.7 to 81.2 °C

| Temperature,°C | $c_p$, J/kgK |
|---|---|
| −18.7 | 656.2 |
| −16.7 | 664.3 |
| −0.7 | 710.6 |
| 1 | 598.9 |
| 5.1 | 671.3 |
| 6.3 | 671.3 |
| 7 | 673.0 |
| 12.1 | 722.2 |
| 13.1 | 704.3 |
| 24.3 | 711.2 |
| 24.8 | 730.9 |
| 25.3 | 724.5 |
| 27.3 | 713.0 |
| 29.7 | 719.9 |
| 30.6 | 715.9 |
| 35.9 | 741.9 |
| 36.4 | 743.1 |
| 43.5 | 770.3 |
| 478 | 764.5 |
| 48.1 | 773.2 |
| 52.5 | 766.8 |
| 53.4 | 771.5 |
| 55.9 | 786.5 |
| 56 | 785.9 |
| 58.7 | 834.0 |
| 64.4 | 769.1 |
| 64.9 | 770.9 |
| 75.9 | 853.1 |
| 76.5 | 858.3 |
| 81.2 | 866.4 |

In order to determine the EER and the COP, the thermodynamic parameters of the circulating medium are defined in the characteristic points of the Linde anti-clockwise cycle (Figs. 5.8 and 5.9). A representation of ideal operation of a refrigeration unit is shown below.

In the above thermodynamic system, the individual EER and COP components can be defined as:

$$Q_K = h_2 - h_3 \quad (5.14)$$

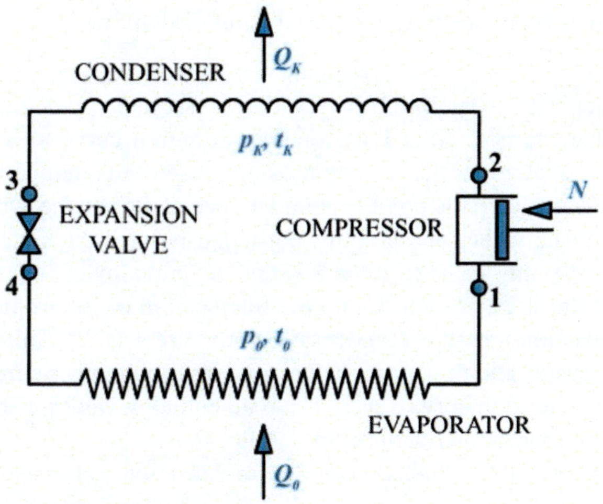


**Fig. 5.8** Single-stage Linde refrigeration cycle

**Fig. 5.9** Single-stage Linde refrigeration cycle in the lg p–h system

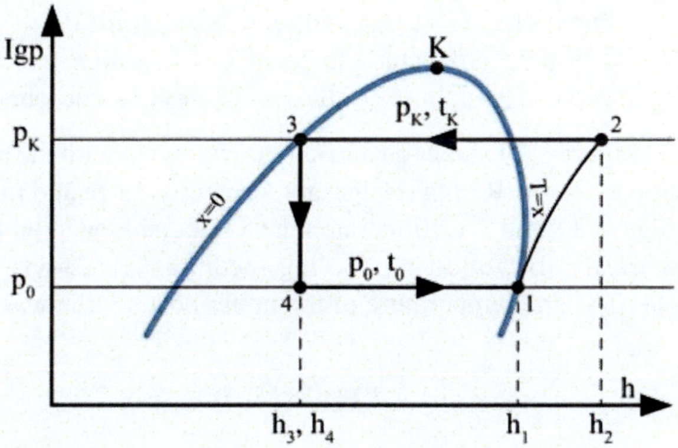


$$Q_o = h_1 - h_4 \tag{5.15}$$

$$L_S = h_2 - h_1 \tag{5.16}$$

In order to determine the thermodynamic values in the characteristic points, the *CoolProp* software package is used to read the following parameters:

h—enthalpy [kJ/kg].

s—entropy [kJ/(kg·K)].

p—pressure [Pa].

*T*—temperature [K].

The CoolProp computational package makes it possible to determine values of the above quantities by specifying a pair of defining parameters.

The calculation process and the required pairs of defining parameters are given below.

1. The first step was to determine evaporation pressure $p_O$ by defining point P1—the first characteristic point lying on the saturation curve x = 1; the second defining parameter was the cooling medium evaporation temperature.
2. The second characteristic point is point P1', which lies on the evaporation pressure isobar. The values defining this particular point is the temperature higher by the superheating value of $\Delta t = 8$ K and the enthalpy of Point P1−h1.
3. The next point is P2, which lies on the intersection of isentropic compression and condensation pressure. Condensation pressure was determined in the same manner as evaporation pressure, using condensation temperature and the position on the saturation curve $x = 1$. The read enthalpy values were corrected by isentropic condensation efficiency $\eta = 0.96$.
4. The next point is P2', which lies on the condensation isobar and the saturation curve x = 1.
5. The next point marks the end of condensation; it lies on the condensation isobar and the saturation curve x = 0.
6. Additional cooling of $\Delta t = -2$ K was used to improve the system efficiency. Parameters $t = t_K + \Delta t$ and $p_K$ define Point P3'.
7. The next characteristic point is P4, which lies on the isenthalpic expansion curve. The defining values were $h_3 = h_4$ and pressure $p_O$.

Figure 5.10 shows an anti-clockwise cycle with the above-presented characteristic points for the R134a refrigerant, currently more and more often used in heat pumps. Figs. 5.11 and 5.12 illustrate the COP dependence on the lower heat source temperature for the following cooling mediums: R1234yf, R134A, R152A, R404A and R410A. The temperature of the upper heat source was assumed at 40 °C (Fig. 5.11)

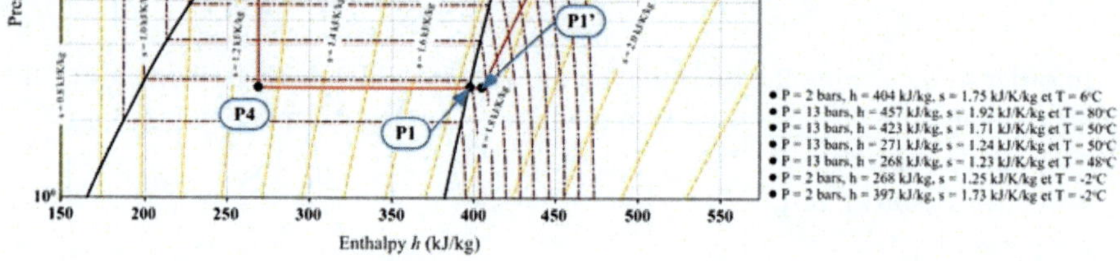


**Fig. 5.10** Single-stage Linde refrigeration cycle in the p–h system for refrigerant R134a

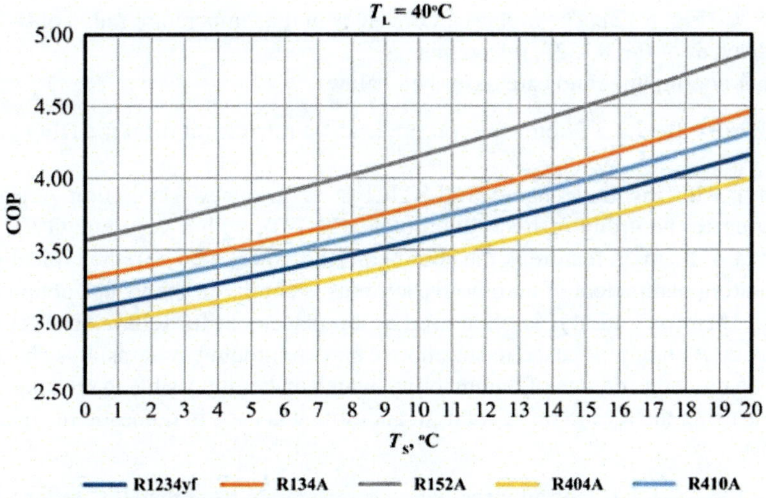


**Fig. 5.11** Dependence of the heat pump COP on the temperature of the lower heat source (at the upper heat source temperature assumed at 40 °C) for different cooling agents: R1234yf, R134A, R152A, R404A and R410A

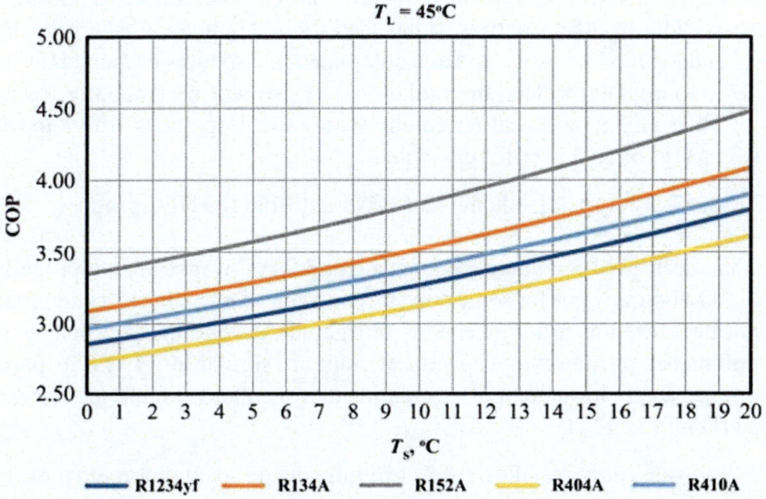


**Fig. 5.12** Dependence of the heat pump COP on the temperature of the lower heat source (at the upper heat source temperature assumed at 45 °C) for different cooling agents: R1234yf, R134A, R152A, R404A and R410A

and 45 °C (Fig. 5.12). The highest COP values in the temperature range of 0–20 °C are achieved for the R152A refrigerant.

Other cooling mediums are described below.

(a) R1234yf—2,3,3,3-tetra-fluoropropylene—a refrigerant from the HFO group

R1234yf with the trade name SOLSTICE® yf is a new-generation refrigerant belonging to the group of hydrofluoroolefins (HFO), with a very low GWP coefficient of 4. Its main feature is the cooling performance, which in the typical range of operating parameters of automotive systems is similar to the values obtained for R134a. Therefore, 1234yf is often used as a replacement for R134a. The R1234yf refrigerant is commonly used in automotive air conditioning systems. It is characterized by a moderate degree of flammability, which makes it possible to retrofit current systems using the R134a refrigerant. In addition, it is used in domestic refrigeration systems.

(b) R134A—1,1,1,2 tetrafluoroethane—a refrigerant from the HFC group

A homogeneous medium. It is one of the most popular synthetic refrigerants, commonly used in air conditioning, mainly automotive, but also in high- and medium-temperature refrigeration; also present in industrial, commercial and domestic applications. Considering its properties, it has also found application as a propellant in technical and medical aerosols, and as a component in mixtures of many HFC refrigerants. Due to climate protection regulations, its use will be gradually reduced in favour of other refrigerants (car air conditioning) and propellants with a low GWP, such as R1234yf or R1234ze for example.

(c) R152A—difluoro-1,1-ethane—a refrigerant from the HFC group

R152A has been used for many years as a component of mixtures, a propellant in aerosols and a foaming gas for some polystyrene foams. So far, due to its flammability, it has not been used as a homogeneous refrigerant, despite the fact that its values of the volumetric refrigerating capacity per unit are similar to R134a. Its particular advantage is the very low potential for creating the greenhouse effect (global warming potential (GWP) = 124).

(d) R404A—a mixture of 1,1,1,2 trifluoroethane, pentafluoroethane, 1,1,1,2 tetrafluoroethane—a refrigerant from the HFC group

A mixture of three agents: 52% R143a, 44% R125 and 4% R134a which is widely used in low-, medium- and high-temperature refrigeration. It is one of the most popular synthetic refrigerants, commonly used in industrial refrigeration: cold stores and freezers, and in commercial refrigeration, e.g. in refrigeration systems in shops. It is also used in air handling units and in refrigerated transport. Due to climate protection regulations, its use will soon be gradually limited in favour of other refrigerants with a low GWP, such as R407F, R448A (Solstice N40), R744. This refrigerant is mainly intended for use in refrigeration and air conditioning systems designed for this particular type of the cooling medium. It can be used in place of withdrawn refrigerants such as R502 or R22 (or their substitutes), but its use requires oil replacement with the POE and replacement of some subassemblies in the installation following appropriate procedures.

(e) R410A—a mixture of difluoromethane, and pentafluoroethane—a refrigerant from the HFC group

A mixture of two agents: 50% R32 and 50% R125. It is used in residential and commercial air conditioning units and heat pumps, e.g. in split air conditioners, ice-water air-cooled chillers and direct evaporation air handling units. As a replacement for R13B1, it can be used in low-temperature systems. It is characterized by a very low temperature slip and can be used in refrigeration systems with flooded evaporators. A cooling agent mainly intended for refrigeration and air conditioning systems designed for this type of refrigerant.

Figures 5.11 and 5.12 briefly present the dependence of the ground source heat pump COP on the temperature of the lower heat source and the upper heat source for the most commonly used refrigerants.

### 5.2.2 A Mathematical Model of RESHeat System with Heat Accumulation in Two Underground Tanks

Figure 5.13 presents a diagram of the RESHeat system with two tanks. Tank 1 is not insulated, whereas Tank 2 has the outer surface insulation.

This section presents the model of RESHeat system. The following system of differential equations is solved:

$$\rho_w c_w \frac{\pi D^2}{4} L \frac{dT_1}{d\tau} = Q_{solPV} - Q_{HP} + Q_{ground,1} \tag{5.17}$$

$$\rho_w c_w \frac{\pi D^2}{4} L \frac{dT_2}{d\tau} = Q_{solSC} - Q_{HP} + Q_{ground,2} \tag{5.18}$$

where

$T_1, T_2$—water temperature in Tank 1 and Tank 2

$\rho_w$—water density

$c_w$—water specific heat

$Q_{solPV}$—heat supplied to the tank by PVT cells

$Q_{solSC}$—heat supplied to the tank by solar collectors

$L$—tank length

$\varphi$—ratio of the amount of thermal energy extracted from Tank1 to the amount of heat extracted from Tank 2

$Q_{HP}$—heat absorbed by the heat pump and transferred to the building

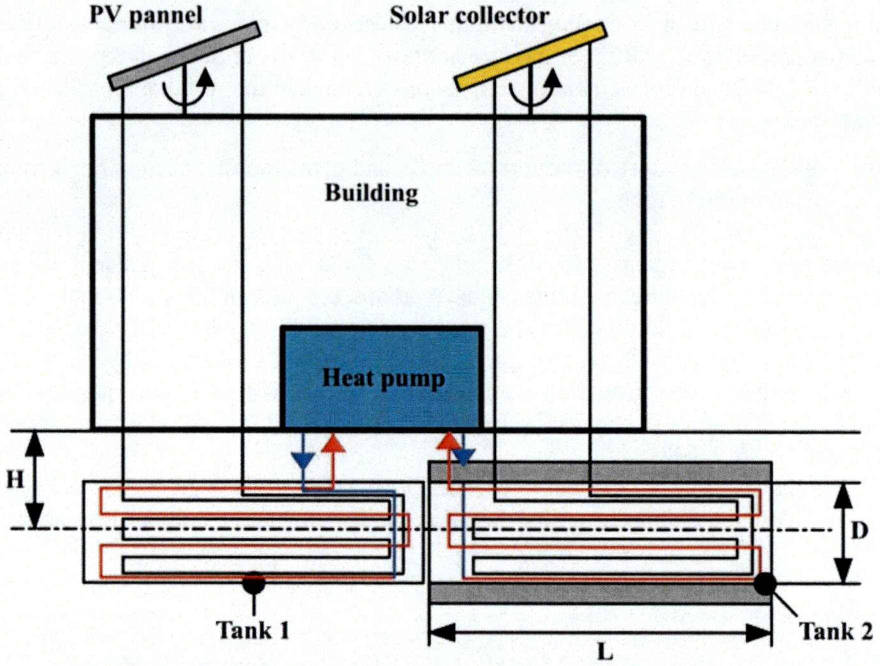


**Fig. 5.13** Scheme of RESHeat system calculation model

$Q_{ground,1}$; $Q_{ground,2}$—heat collected from the ground by Tank 1 and Tank 2, respectively

$D$—tank outer diameter.

The following model is adopted for the description of heat accumulation in the ground:

$$Q_{solPV} = \mathrm{q}A_{PV}\,\eta_{sc} \tag{5.19}$$

$$Q_{solSC} = \mathrm{q}A_{sc}\,\eta_{sc} \tag{5.20}$$

$$Q_{HP} = Q_b\left(1 - \frac{1}{COP}\right) \tag{5.21}$$

$$Q_b = UV_b(T_{set} - T_{ext}) \tag{5.22}$$

$$COP = \frac{Q_O + L_S}{L_S} \tag{5.23}$$

$$Q_{ground,1} = U_1\pi DL\left(T_g - T_1\right) \tag{5.24}$$

$$Q_{ground,2} = U_2 \pi DL(T_g - T_2) \tag{5.25}$$

where $U_1 = 250$ W/(m$^2$ K) is the water-ground heat transfer coefficient for the non-insulated tank, and $U_2 = 0.5$ W/(m$^2$ K) is the water-ground heat transfer coefficient for the insulated tank.

$$\rho c \frac{\partial T_g}{\partial \tau} = k \frac{\partial^2 T_g}{\partial x^2} + k \frac{\partial^2 T_g}{\partial y^2} \tag{5.26}$$

The boundary conditions take the following form (according to Fig. 5.14):

$$\left.\frac{\partial T_g}{\partial x}\right]_{x=-w} = 0 \tag{5.27}$$

$$\left.\frac{\partial T_g}{\partial y}\right]_{y=-H_t} = 0 \tag{5.28}$$

$$\left.-k\frac{\partial T_g}{\partial y}\right]_{y=-H_1} = U_g(T_g - T_{ext}) \tag{5.29}$$

For Tank 1 internal wall:

$$\left.-k\frac{\partial T_g}{\partial r}\right]_{r=r_i} = U_1(T_g - T_1) \tag{5.30}$$

For Tank 2 internal wall:

$$\left.-k\frac{\partial T_g}{\partial r}\right]_{r=r_i} = U_2(T_g - T_2) \tag{5.31}$$

The system parameters adopted for the calculations are presented in Table 5.3.

The monthly averaged solar irradiation $q$ considered in the calculation is given in Table 5.4. The external (ambient) temperature $T_{\text{ext}}$ is calculated by using Kusuda and Achenbach (1965) model for ground surface level.

Figure 5.14 shows the numerical mesh used in the calculations.

### 5.2.3 *Discretization of Energy Equation—Finite Volume Method*

The energy balance equations were created using the finite volume method (FVM). The method is applied to solve heat conduction problems and is implemented, among others, in the FLUENT commercial code for CFD (computational fluid dynamics) calculations. It can be used to determine the temperature field for solids characterized

**Table. 5.3** Parameters of the model of the RESHeat system with two accumulation tanks

| Parameter of the model | Description | Value |
|---|---|---|
| $D$ | Accumulation tank diameter | 2 m, 3 m |
| $L$ | Tank length | 10 m |
| $W$ | Ground width | 10 m |
| $H_t$ | Ground depth | 10 m |
| $H_1$ | Tank burial depth | 3.5 m |
| $U_1$ | Water-ground overall heat transfer coefficient in the non-insulated tank | 250 W/(m$^2$ K) |
| $U_2$ | Water-ground overall heat transfer coefficient in the insulated tank | 0.05 W/(m$^2$ K) |
| $U_g$ | Air-to-insulated ground overall heat transfer coefficient (external surface) | 0.7 W/(m$^2$ K) |
| $T_r = UV_b$ | Thermal transmittance of the building | 960 W/K |
| $\rho_w$ | Water density | 1000 kg/m$^3$ |
| $c_w$ | Water specific heat | 4180 J/(kg K) |
| $A_{SC}$ | Surface area of solar collectors | 54.4 m$^2$ |
| $A_{PV}$ | Surface area of photovoltaic panels | 259.2 m$^2$ |
| $\eta_{PV}$ | Thermal efficiency of PV panels | 0.45 |
| $\eta_{SC}$ | Thermal efficiency of solar collectors | 0.75 |
| $T_{load}$ | Upper heat source temperature | 40 °C |
| $T_g(\tau = 0)$ | Ground initial temperature | 20 °C, 7 °C |
| $T_1(\tau = 0)$ | Initial temperature of water in the non-insulated tank | 20 °C |
| $T_2(\tau = 0)$ | Initial temperature of water in the insulated tank | 20 °C |

**Table. 5.4** Solar radiation intensity averaged over one month and number of hours of sunshine per day

| Month | Averaged heat flux, ($q$ W/m$^2$) | Number of hours of sunshine per day |
|---|---|---|
| January | 400 | 2.8 |
| February | 400 | 3.6 |
| March | 500 | 4.8 |
| April | 600 | 7.0 |
| May | 800 | 7.7 |
| June | 800 | 7.9 |
| July | 900 | 8.8 |
| August | 900 | 7.5 |
| September | 700 | 6.1 |
| October | 500 | 4.8 |
| November | 400 | 3.4 |
| December | 400 | 2.5 |

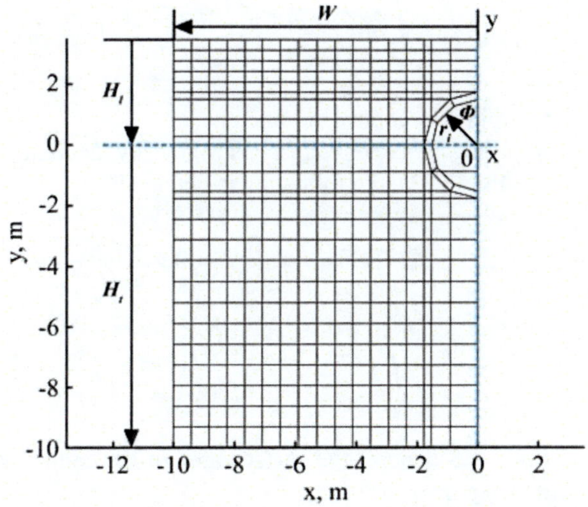


**Fig. 5.14** Numerical mesh adopted for the temperature distribution calculations

by both irregular shapes and different thermal properties. The first step in the method is to formulate an integral form of the energy conservation equation. Next, the equations are discretized, which means that they are written in a form appropriate for a specific control volume. As a result, energy conservation equations are obtained in a global form. The second step is to divide the entire area into a set of non-overlapping control volumes. The energy balance equations for the control volumes are used to determine the temperature in the control volume internal nodes. By determining such equations for all volumes, a system of differential equations is obtained, which is then solved using numerical methods for transient problems (the Explicit method, the Implicit method, the Crank–Nicholson method). In the case under analysis, the control volume method was used to discretize the equation defining heat conduction in the ground.

The theoretical basis of the finite volume method is presented based on (Taler and Duda, 2006).

If the thickness of the analysed area is $\Delta$ and the thermal properties are known (specific heat $c$, density ρ, heat conductivity coefficient $k$) and if also known is the heat flux of the volumetric heat sources $\dot{q}_v$, the general heat conduction equation can be written as:

$$c(T) \cdot \rho(T) \frac{\partial T}{\partial t} = -\nabla \cdot \dot{q} + \dot{q}_v \tag{5.32}$$

where: $T$—temperature, $t$—time, ∇—nabla operator.

The analysed area is divided into control volumes with dimensions $\Delta x$, $\Delta y$ and ∇ in the Cartesian coordinate system or $\Delta r$, $\Delta \varphi$ and Δ in the cylindrical system of coordinates.

By integrating Eq. (5.32) over the control volume, a general equation is obtained for a single control volume and expressed as:

$$\int_{CV} c(T) \cdot \rho(T) \frac{\partial T}{\partial t} dV = -\int_{CV} \nabla \cdot \dot{q} dV + \int_{CV} \dot{q}_v dV \tag{5.33}$$

where: *CV*—control volume.

Applying the Green-Gauss-Ostrogradsky theorem to the first term of the left-hand side of Eq. (5.33), the following is obtained:

$$\int_{CV} \nabla \cdot q dV = \int_{CS} q \cdot n dS \tag{5.34}$$

$$\int_{CV} c(T) \cdot \rho(T) \frac{\partial T}{\partial t} dV = -\int_{CS} q \cdot n dS + \int_{CV} \dot{q}_v dV \tag{5.35}$$

where CS denotes the surface area of the control volumes and $n$ is the external normal vector.

The normal vector enables determination of a relation in the form of Eq. (5.35); according to the equation, it can be seen that when heat flows into the control volume, the vector of heat flux $q$ is directed towards the control volume inside. The angle between vectors $n$ and $q$ is 180°.

$$n \cdot q = 1 \cdot |q| \cos(n, q) = q_n \tag{5.36}$$

If $\Delta V$ denotes a single-cell control volume, then, after reduction, Eq. (5.35) takes the following form:

$$\int_{CV} c(T) \cdot \rho(T) \frac{\partial T}{\partial t} dV \cong \Delta V_c(T_p) \rho(T_p) \frac{dT_p}{dt} \tag{5.37}$$

$$-\int_{CS} q \cdot n dS = \sum_{i=1}^{4} \dot{Q}_i \tag{5.38}$$

$$\int_{CV} \dot{q}_v dV = \Delta V \, \dot{q_v}(T_p) \tag{5.39}$$

where $\dot{Q}_i$ is the heat flux delivered to an adjacent cell.

Substituting Eqs. (5.36–5.39) in Eq. (5.34) results in the general energy balance equation, which can be written as:

$$\Delta V_c(T_p) \rho(T_p) \frac{dT_p}{dt} = \sum_{i=1}^{4} \dot{Q}_i + \Delta V \, \dot{q_v}(T_p) \tag{5.40}$$

The general energy balance Eq. (5.40) can be described in detail in the Cartesian and in the cylindrical system of coordinates.

### 5.2.4 General Energy Balance Equations in the Cartesian Coordinate System

Figure 5.15 shows the analysed area division into control volumes. For a single cell, it is assumed that $\Delta V = (\Delta x)\cdot(\Delta y)\ \Delta$. Where $\Delta$ denotes thickness and is assumed to be equal to 1 m. Heat flows from adjacent cells of the control volume from nodes W, N, E and S to the central node – P.

The heat flow rate for the general case presented in Fig. 5.15 can be written as:

$$\dot{Q}_{W-P} = (\Delta y) \cdot \Delta \cdot \frac{k(T_W) + k(T_P)}{2} \cdot \frac{T_W - T_P}{\Delta x} \tag{5.41}$$

$$\dot{Q}_{N-P} = (\Delta x) \cdot \Delta \cdot \frac{k(T_N) + k(T_P)}{2} \cdot \frac{T_N - T_P}{\Delta y} \tag{5.42}$$

$$\dot{Q}_{E-P} = (\Delta y) \cdot \Delta \cdot \frac{k(T_E) + k(T_P)}{2} \cdot \frac{T_E - T_P}{\Delta x} \tag{5.43}$$

$$\dot{Q}_{S-P} = (\Delta x) \cdot \Delta \cdot \frac{k(T_S) + k(T_P)}{2} \cdot \frac{T_S - T_P}{\Delta y} \tag{5.44}$$

where: $\Delta x$ and $\Delta y$ denote the control volume dimensions.

Substituting Eqs. (5.40–5.44) in the general Eq. (5.40), the result is as follows:

$$\begin{aligned}&(\Delta x) \cdot (\Delta y) \cdot \Delta \cdot c(T_p)\rho(T_p)\frac{dT_p}{dt}\\&= (\Delta y) \cdot \Delta \cdot \frac{k(T_W) + k(T_P)}{2} \cdot \frac{T_W - T_P}{\Delta x} + (\Delta x) \cdot \Delta\end{aligned}$$

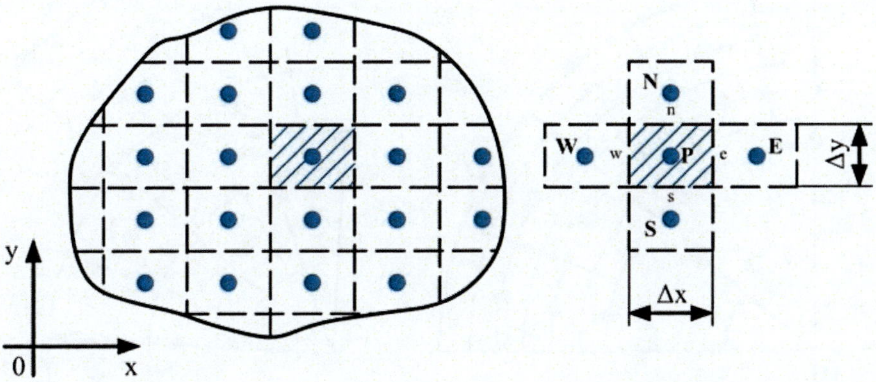


**Fig. 5.15** Two-dimensional region divided into control volumes in the Cartesian system–redrawn based on (Taler and Duda, 2006)

$$\cdot\frac{k(T_N)+k(T_P)}{2}\cdot\frac{T_N-T_P}{\Delta y}+(\Delta y)\cdot\Delta$$
$$\cdot\frac{k(T_E)+k(T_P)}{2}\cdot\frac{T_E-T_P}{\Delta x}+(\Delta x)\cdot\Delta$$
$$\cdot\frac{k(T_S)+k(T_P)}{2}\cdot\frac{T_S-T_P}{\Delta y}+(\Delta x)\cdot(\Delta y)\cdot\Delta\cdot\dot{q}_v \tag{5.45}$$

After transformation Eq. (5.44) takes the following form:

$$\frac{dT_p}{dt}=\alpha_P\left[\frac{k_W+k_P}{2\cdot k_P}\cdot\frac{T_W-T_P}{(\Delta x)^2}+\frac{k_N+k_P}{2\cdot k_P}\cdot\frac{T_N-T_P}{(\Delta y)^2}+\frac{k_E+k_P}{2\cdot k_P}\cdot\frac{T_E-T_P}{(\Delta x)^2}+\frac{k_S+k_P}{2\cdot k_P}\cdot\frac{T_S-T_P}{(\Delta y)^2}\right]+\frac{\dot{q}_v}{c_P\cdot\rho_P} \tag{5.46}$$

where: $k_p=k(T_p)$, $c_p=c(T_p)$, $\rho_p=\rho(T_p)$, $\alpha_p=k_p/(c_p\cdot\rho_p)$.

In steady-state conditions, expression $dT_P/dt=0$. For an identical $\Delta x=\Delta y$ mesh, and for constant and temperature-independent thermal properties and power of the heat sources, expression (5.46) takes the following form:

$$T_W+T_N+T_E+T_S-4T_P+\frac{\dot{q}_v(\Delta x)^2}{k}=0 \tag{5.47}$$

Figure 5.16 presents the two-dimensional region divided into control volumes in the cylindrical system.

The energy balance equation in the cylindrical system can be described using Eqs. (5.48–5.52) presented below:

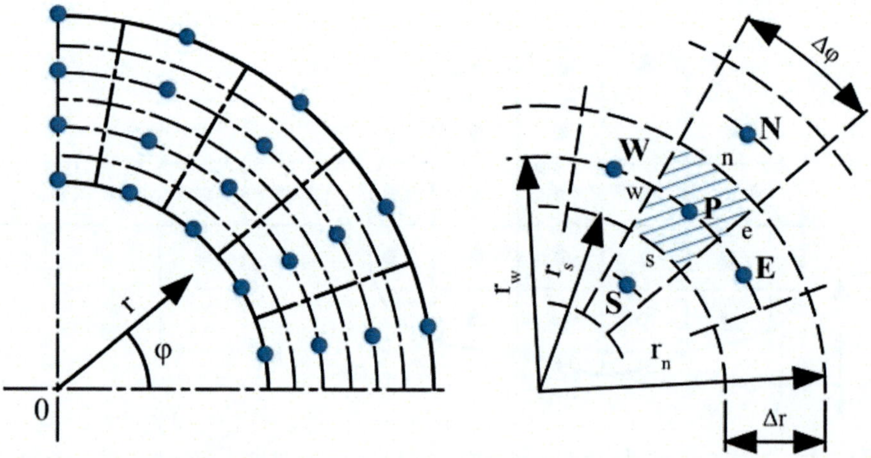


**Fig. 5.16** Two-dimensional region divided into control volumes in the cylindrical system–redrawn based on (Taler and Duda, 2006)

$$\Delta V = \pi \cdot \left(r_n^2 - r_s^2\right) \cdot \frac{\Delta\varphi}{2\pi} \cdot \Delta = \frac{\left(r_n^2 - r_s^2\right) \cdot \Delta\varphi}{2} \cdot \Delta \tag{5.48}$$

$$\dot{Q}_{W-P} = (\Delta r) \cdot \Delta \cdot \frac{k_W + k_P}{2} \cdot \frac{T_W - T_P}{r_w \cdot (\Delta\varphi)} \tag{5.49}$$

$$\dot{Q}_{N-P} = r_n \cdot (\Delta\varphi) \cdot \Delta \cdot \frac{k_N + k_P}{2} \cdot \frac{T_N - T_P}{\Delta r} \tag{5.50}$$

$$\dot{Q}_{E-P} = (\Delta r) \cdot \Delta \cdot \frac{k_E + k_P}{2} \cdot \frac{T_E - T_P}{r_w \cdot (\Delta\varphi)} \tag{5.51}$$

$$\dot{Q}_{S-P} = r_n \cdot (\Delta\varphi) \cdot \frac{k_S + k_P}{2} \cdot \frac{T_S - T_P}{\Delta r} \tag{5.52}$$

Substituting Eqs. (5.48–5.52) in Eq. (5.40), the result is as follows:

$$\begin{aligned}\frac{dT_p}{dt} = \frac{2 \cdot \alpha_P}{\left(r_n^2 - r_s^2\right) \cdot \Delta\varphi} \Bigg[ & \frac{k_W + k_P}{2k_P} \cdot \frac{\Delta r}{r_w \cdot (\Delta\varphi)} \cdot (T_W - T_P) + \frac{k_N + k_P}{2k_P} \\ & \cdot \frac{r_n \cdot (\Delta\varphi)}{\Delta r} \cdot (T_N - T_P) + \frac{k_E + k_P}{2k_P} \cdot \frac{\Delta r}{r_w \cdot (\Delta\varphi)} \\ & \cdot (T_E - T_P) + \frac{k_S + k_P}{2k_P} \cdot \frac{r_S \cdot (\Delta\varphi)}{\Delta r} \cdot (T_S - T_P) + \frac{q_v}{c_p \cdot \rho_p} \Bigg]\end{aligned} \tag{5.53}$$

where: $\alpha_p = k_p/(c_p \cdot \rho_p)$ is thermal diffusivity. For each node the heat balance equation is formulated based on the schemes given by Eq. (5.46) and (5.53). Then, the system of Ordinary Differential Equations is solved using the Euler method.

### 5.2.5 Results and Discussion

The calculations were carried out for two different diameters of the accumulation tanks: $D = 2$ m and $D = 3$ m. May was assumed as the month of the system installation. The calculations were carried out for two different initial temperatures of the ground: $T_g$ ($\tau = 0$) = 20 °C and 7 °C.

Figure 5.17 presents the curves illustrating the building cumulated demand for thermal energy $Q_{dem}$ and the amount of heat produced and delivered to the building by the heat pump using the accumulation tanks $Q_{prod}$. If the water temperature in the accumulation tanks is lower than 3 °C, the heat pump uses the thermal energy from the ground heat exchanger. The analysed case concerns the heating season and it can be seen clearly that with the use of the heat accumulation unit in the ground it is possible to cover a significant part of the building demand for heat during that time. The calculations were carried out for the ground initial temperature $T_g$ ($\tau = 0$) = 20 °C.

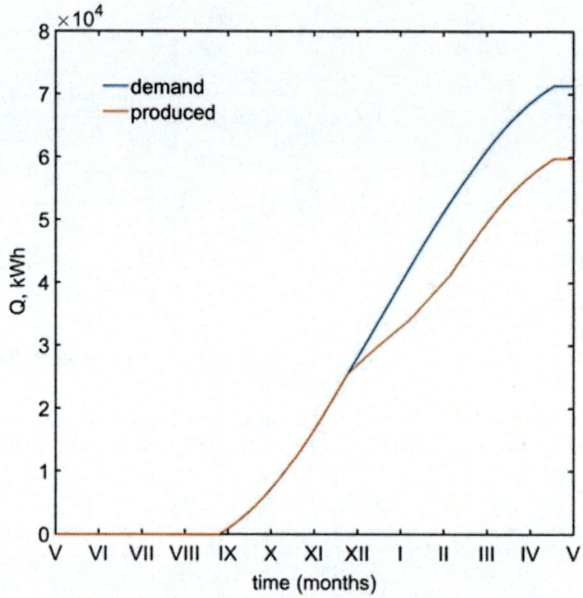


**Fig. 5.17** Building cumulated demand for heat $Q_{dem}$ and the amount of heat $Q_{prod}$ supplied to the building by the heat pump in the first year of the system operation

Figure 5.18 shows the temperature distribution in the accumulation tanks. Temperature $T_2$ is the temperature of water in the insulated tank, whereas temperature $T_1$ is the temperature of water in the non-insulated tank. The dashed line represents the outdoor (ambient) temperature. During the first days of the system operation, the heat

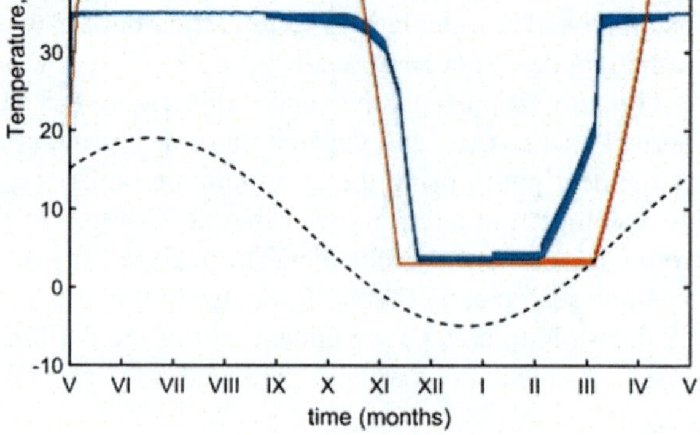


**Fig. 5.18** History of the temperature of water in the non-insulated tank $T_1$, in the insulated tank $T_2$ and outdoor (ambient) temperature $T_{ext}$ in the first year of the system operation

from the PVT cells and from the solar collectors is supplied to the heat accumulation unit. The water temperature in the non-insulated tank is 35 °C maximum, while in the insulated tank it is 65 °C. During the heating season, heat is first extracted from the non-insulated tank and then from the high-temperature (insulated) tank.

The *rat* coefficient describing the coverage of the heat demand using an accumulation unit is 85%. The following parameters of the ground are assumed in the calculations: $k = 1$ W/(m K), $\rho = 1600$ kg/m$^3$, $c_p = 1400$ J/(kg K).

During the summer period, the non-insulated tank is heated to the temperature of 35 °C. Next it gives up heat to the ground. During the heating season, the heat accumulated in the insulated tank is used first, followed by the heat from the non-insulated tank. If the water temperature in the accumulation tank drops below 3 °C, ground heat exchangers are used in the proposed heating system. The system under analysis makes it possible to fully cover the building demand for heat till mid-November. In the December-February period, the heat accumulated in the ground and the excess heat from solar collectors enable effective coverage of a significant part of the thermal energy needed to heat the building.

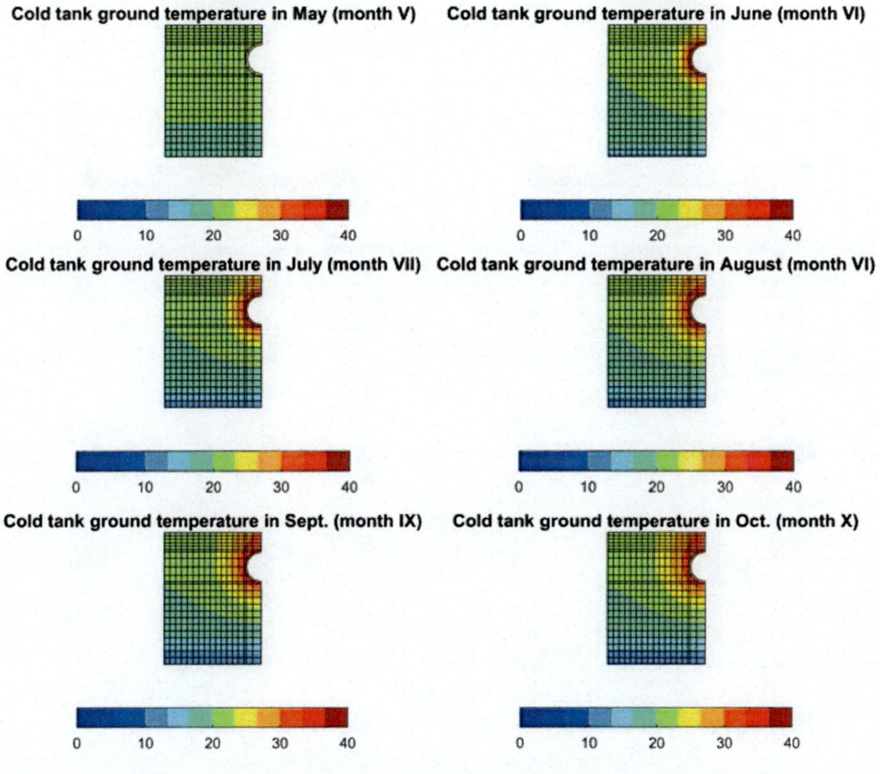


**Fig. 5.19** Ground temperature distribution around the non-insulated tank in the period from May to October (the last day of the month); accumulation tank diameter: $D = 3$ m

Figure 5.19 shows the ground temperature distribution around the non-insulated tank during the period from May to October. The accumulation tank begins to heat up in May; in June the water in the tank reaches the temperature of 35 °C, which is maintained for the following months. The ground temperature around the tank also begins to rise to over 30 °C and remains at this level until October.

Figure 5.20 shows the ground temperature distribution around the insulated tank; the ground temperature around the tank totals 25 °C and changes only slightly in the May–October period.

Figure 5.21 shows the ground temperature distribution around the non-insulated tank during the November–April period (heating season). In November, the ground temperature around the non-insulated tank is high and exceeds 30 °C. Owing to that, heat from the ground is efficiently transferred to the tank. In the period of December-January, due to high consumption of thermal energy, the water temperature in the accumulation tank drops below 3 °C, so additional heat extraction is needed from the boreholes.

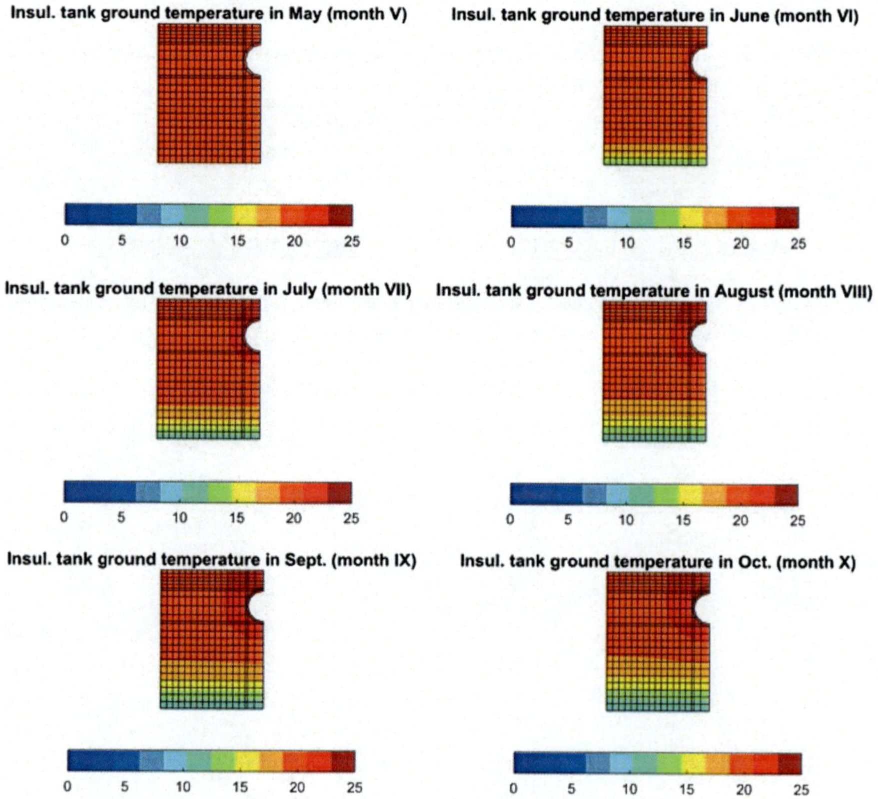


**Fig. 5.20** Ground temperature distribution around the insulated tank in the period from May to October (the last day of the month); accumulation tank diameter: $D = 3$ m

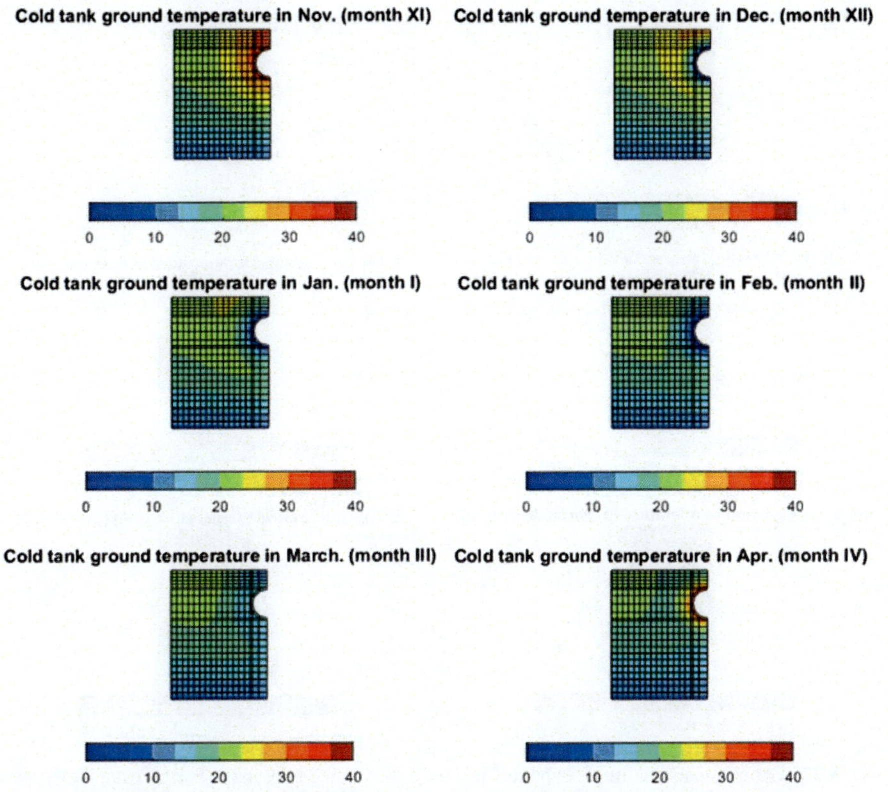


**Fig. 5.21** Ground temperature distribution around the non-insulated tank in the period from November to April (the last day of the month); accumulation tank diameter: $D = 3$ m

Figure 5.22 shows the ground temperature distribution around the insulated tank during the November–April period. Owing to the system insulation, the ground temperature around the tank is up to 20 °C.

Similar temperature distributions are presented for the tank with the diameter $D = 2$ m for the period from May to October (Figs. 5.23 and 5.24 for the non-insulated and the insulated tank, respectively). Figure 5.25 shows the temperature distributions for the non-insulated tank during the November–April period, whereas Fig. 5.26 presents the same for the insulated tank.

Figure 5.27 illustrates the impact of the tank diameter on changes in the water temperature in the non-insulated and in the insulated tank ($T_1$) and ($T_2$), respectively. The analysis of the impact was performed for two different diameters of the tank: 2 and 3 m, respectively. In the tank with the smaller diameter ($D = 2$ m) water heats up much faster compared to the tank with the bigger diameter ($D = 3$ m). However, due to the accumulation tank smaller capacity, the time it takes for the tank to cool down is also much shorter.

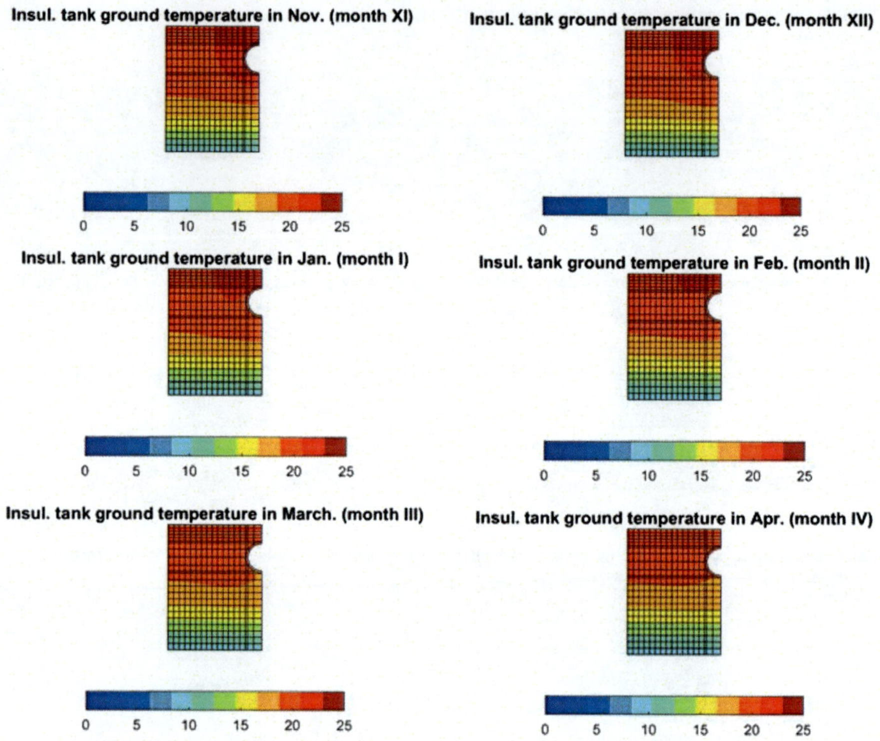


**Fig. 5.22** Ground temperature distribution around the insulated tank in the period from November to April (the last day of the month); accumulation tank diameter: $D = 3$ m

Figure 5.28 shows the building cumulated demand for heat $Q_{dem}$ and the amount of heat $Q_{prod}$ supplied to the building by the heat pump in the first year of the system operation for the tank diameters $D = 2$ m and $D = 3$ m. The system with accumulation tanks with diameter $D = 2$ m covers 77% of the analysed building demand for thermal energy, i.e. by 9% less than the system with two tanks with diameter $D = 3$ m.

The calculation results presented in Figs. 5.17, 5.18, 5.19, 5.20, 5.21, 5.22, 5.23, 5.24, 5.25, 5.26, 5.27, 5.28 were obtained for the ground initial temperature $T_g$ ($\tau = 0) = 20$ °C. Calculations were performed next for the ground initial temperature $T_g$ ($\tau = 0) = 7$ °C to check the temperature impact on the degree of coverage of the system demand for thermal energy.

The heat demand coverage degree is defined as:

$$\psi = \frac{\int_o^{t_{year}} Q_{prod} d\tau}{\int_o^{t_{year}} Q_b d\tau} \tag{5.54}$$

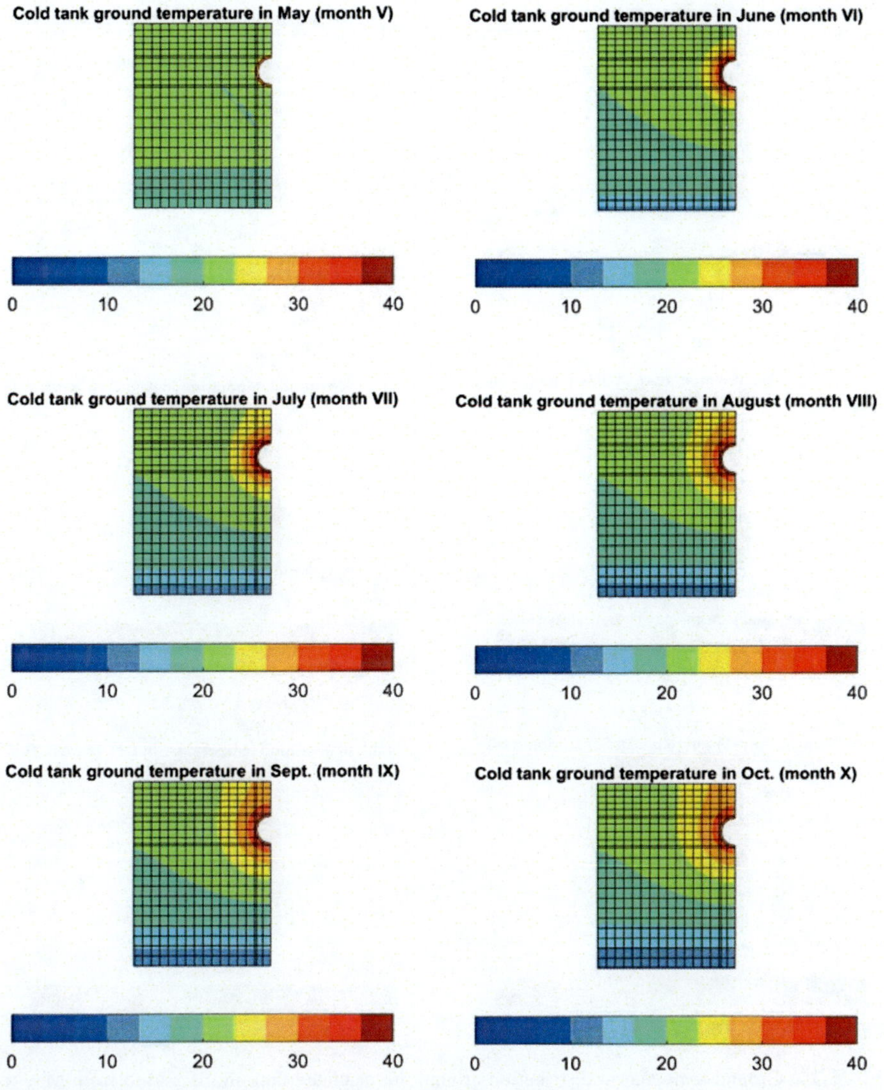


**Fig. 5.23** Ground temperature distribution around the non-insulated tank in the period from May to October (the last day of the month); accumulation tank diameter: $D = 2$ m

where $Q_{prod}$ is a variable depending on the building demand for thermal energy (the system thermal power given in KW) and $t_{year}$ is the period of one year of the heat pump operation.

Figure 5.29 shows the ground temperature distribution around the non-insulated tank during the May–October period. Analysing Fig. 5.29, it can be seen clearly that in the September–October period the ground temperature around the accumulation tank exceeds 20 °C.

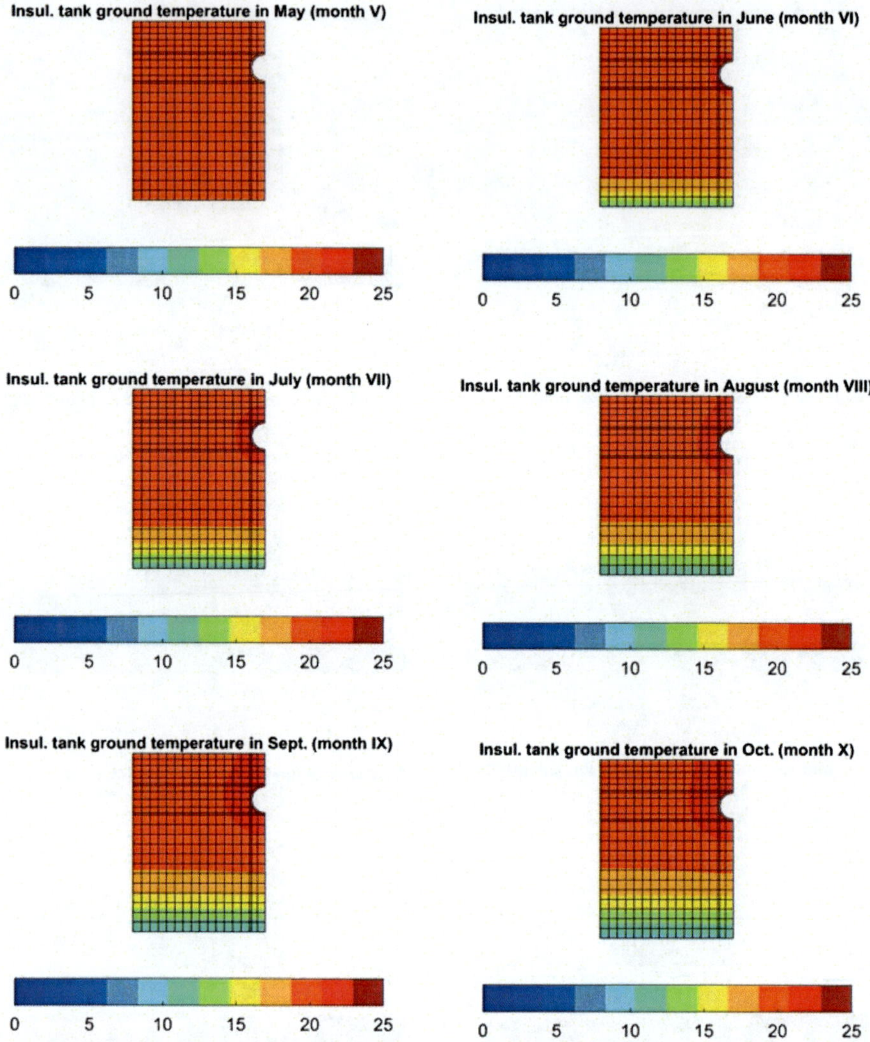


**Fig. 5.24** Ground temperature distribution around the insulated tank in the period from May to October (the last day of the month); accumulation tank diameter: $D = 2$ m

Figure 5.30 shows the ground temperature distribution around the insulated tank. Considering that the heat transfer between the tank and the ground is almost none, the ground temperature in September and October (after the summer season) exceeds 10 °C.

Figure 5.31 shows the temperature distribution for the non-insulated tank in the heating season (November–December). It can be observed that heat is transferred from the ground to the tank. The heat exchange between the ground and the tank

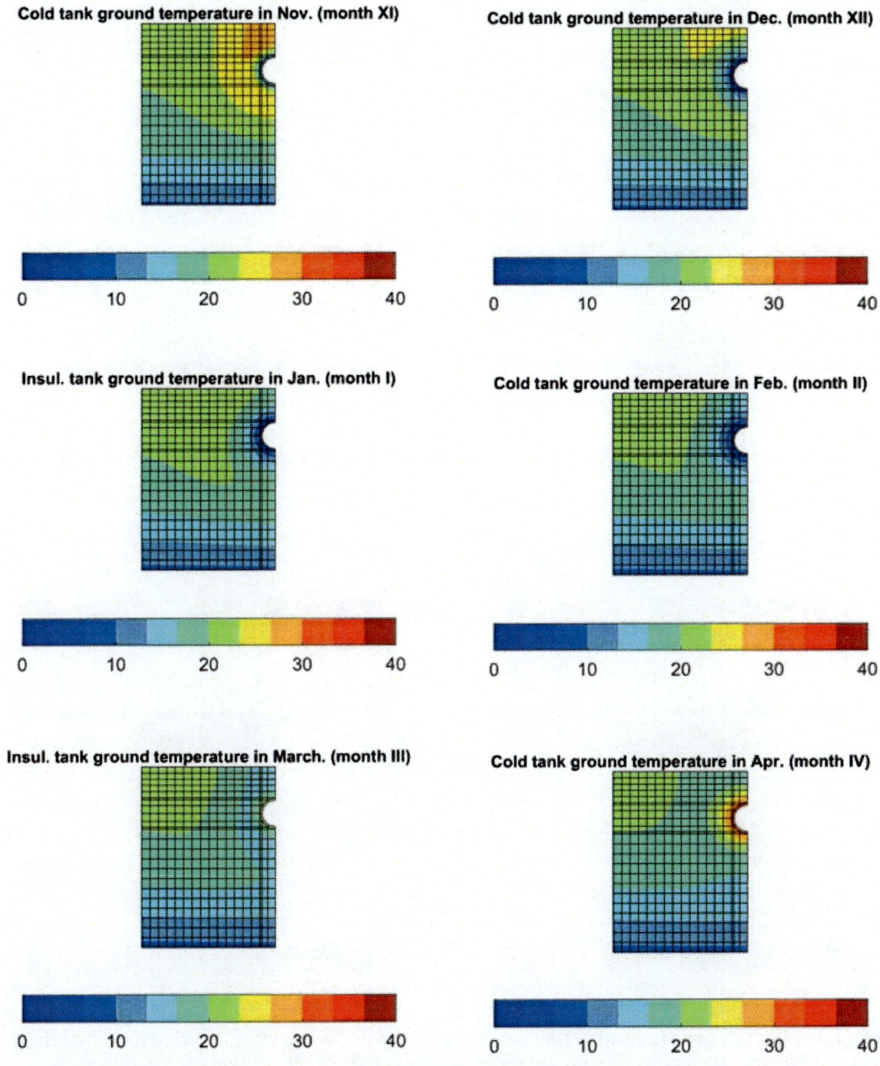


**Fig. 5.25** Ground temperature distribution around the non-insulated tank in the period from November to April (the last day of the month); accumulation tank diameter: $D = 2$ m

is much less intensive in the January–February period due to the small temperature gradient between the ground and the water in the tank.

Figure 5.32 shows the ground temperature distribution for the insulated tank during the November–April period. Because both the tank and the ground surface are insulated, no significant differences occur in the ground temperature in the November-February period.

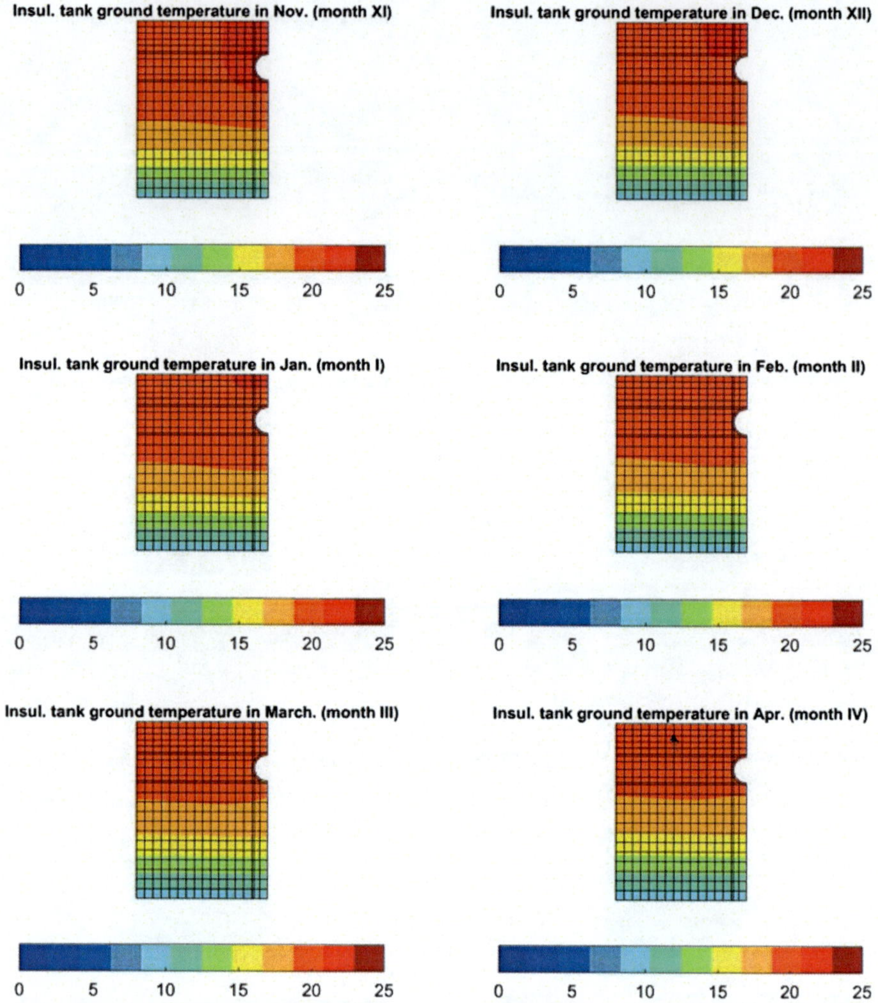


**Fig. 5.26** Ground temperature distribution around the insulated tank in the period from November to April (the last day of the month); accumulation tank diameter: $D = 2$ m

Figure 5.33 presents the amount of thermal energy delivered to the building and the water temperature distribution in the non-insulated tank ($T_1$) and in the insulated tank ($T_2$). The calculated degree of the coverage of the thermal energy demand for the ground initial temperature $T_g$ $(\tau = 0) = 7$ °C is $\Psi = 75.9\%$. This is a value by 1.77% lower compared to the ground initial temperature of 20 °C ($\Psi = 77.67\%$).

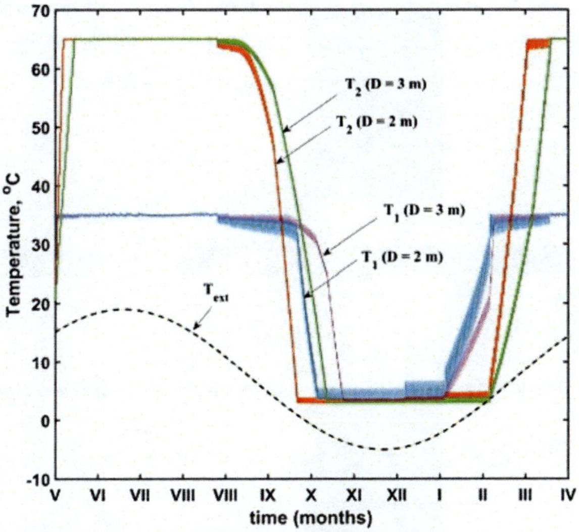


**Fig. 5.27** Impact of the diameter of accumulation tanks on changes in the temperature of water in the non-insulated and in the insulated tank—($T_1$) and ($T_2$), respectively; tank diameters under analysis: $D = 2$ m and $D = 3$ m

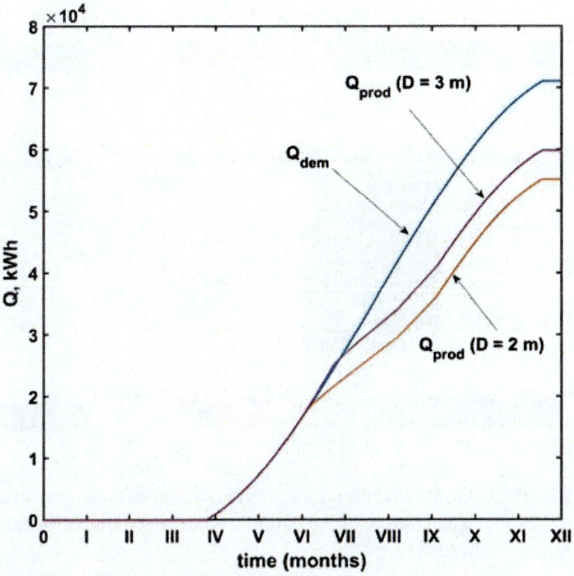


**Fig. 5.28** Building cumulated demand for heat $Q_{dem}$ and the amount of heat $Q_{prod}$ supplied to the building by the heat pump in the first year of the system operation for the tank diameters $D = 2$ and $D = 3$ m

### *5.2.6 Analysis of the Impact of the Ground Thermophysical Properties on the Performance of the Heat Accumulation System*

Figure 5.34 shows the impact of the ground parameters (thermal conductivity, specific

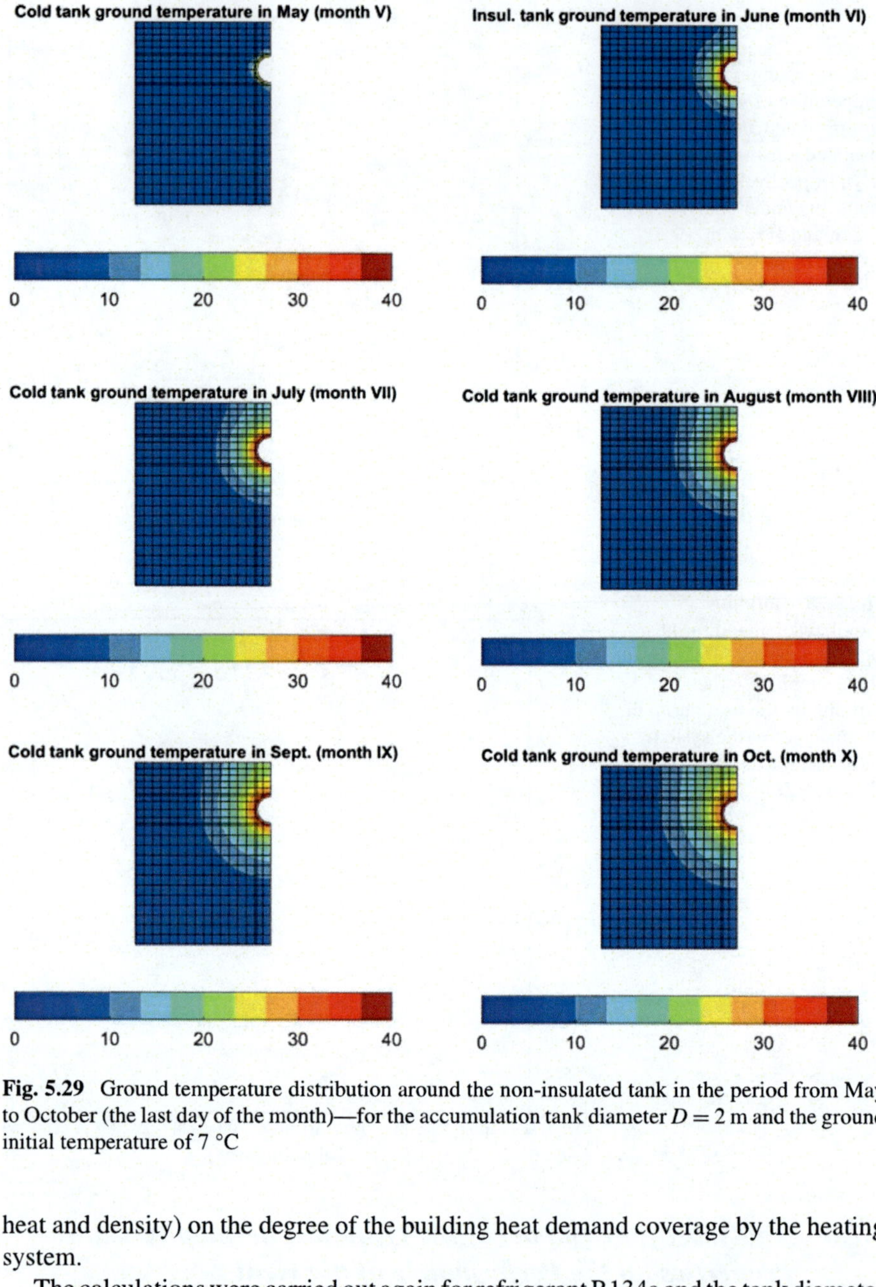


**Fig. 5.29** Ground temperature distribution around the non-insulated tank in the period from May to October (the last day of the month)—for the accumulation tank diameter $D = 2$ m and the ground initial temperature of 7 °C

heat and density) on the degree of the building heat demand coverage by the heating system.

The calculations were carried out again for refrigerant R134a and the tank diameter $D = 2$ m. The efficiency of the proposed system increases with a rise in specific heat, thermal conductivity and density of the ground. For dry soil with thermal conductivity $k = 0.8$ W/(m K), density $\rho = 1200$ kg/m$^3$ and specific heat $c_p = 1000$ J/(kg K), the degree of coverage of the heat demand is 77%. If the ground is moist and thermal

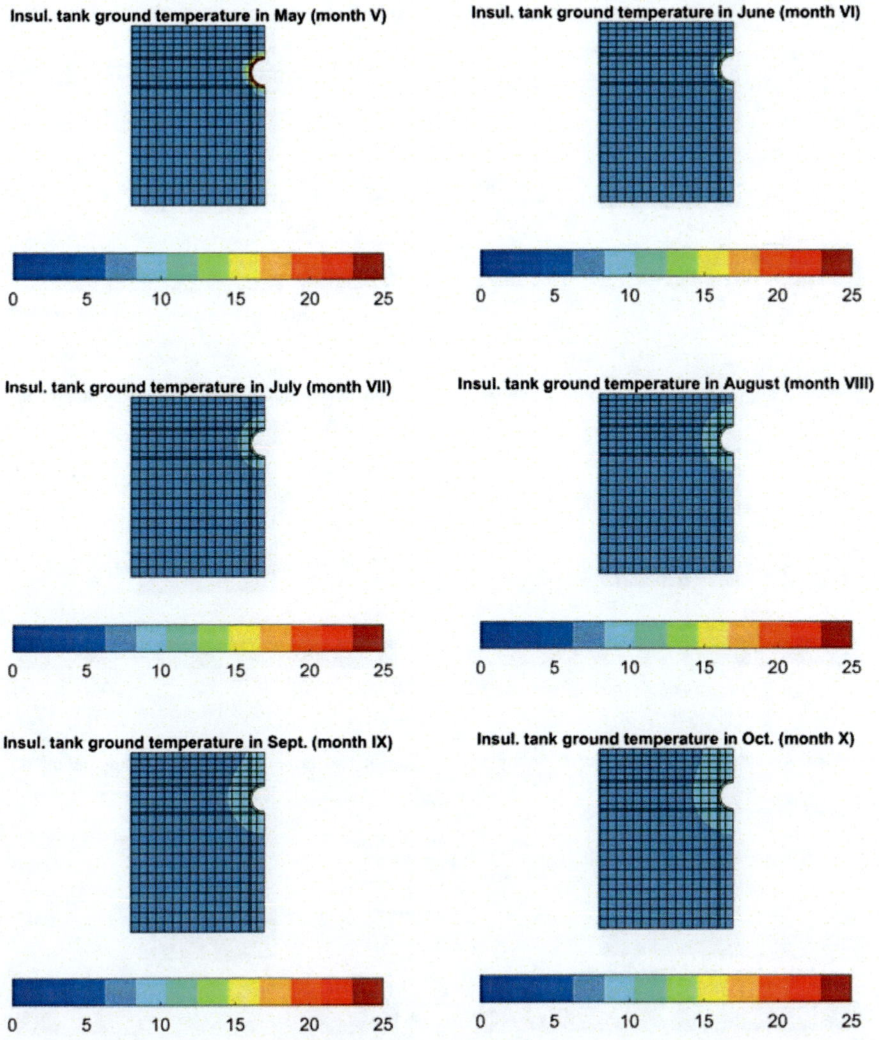


**Fig. 5.30** Ground temperature distribution around the insulated tank in the period from May to October (the last day of the month)—for the accumulation tank diameter $D = 2$ m and the ground initial temperature of 7 °C

conductivity is $k = 2.5$ W/(m K), the building thermal energy demand is covered in 81.5% (for density $\rho = 1800$ kg/m$^3$ and specific heat $c_p = 1600$ J/(kg K)).

In the case of wet soil, with good heat conduction and heat accumulation capabilities, the efficiency of the analysed system is by 4.5% higher compared to dry soil.

Figure 5.35 shows the dependence of the annual average COP of the heat pump on the physical properties of the ground for the system under analysis. Only the unit of

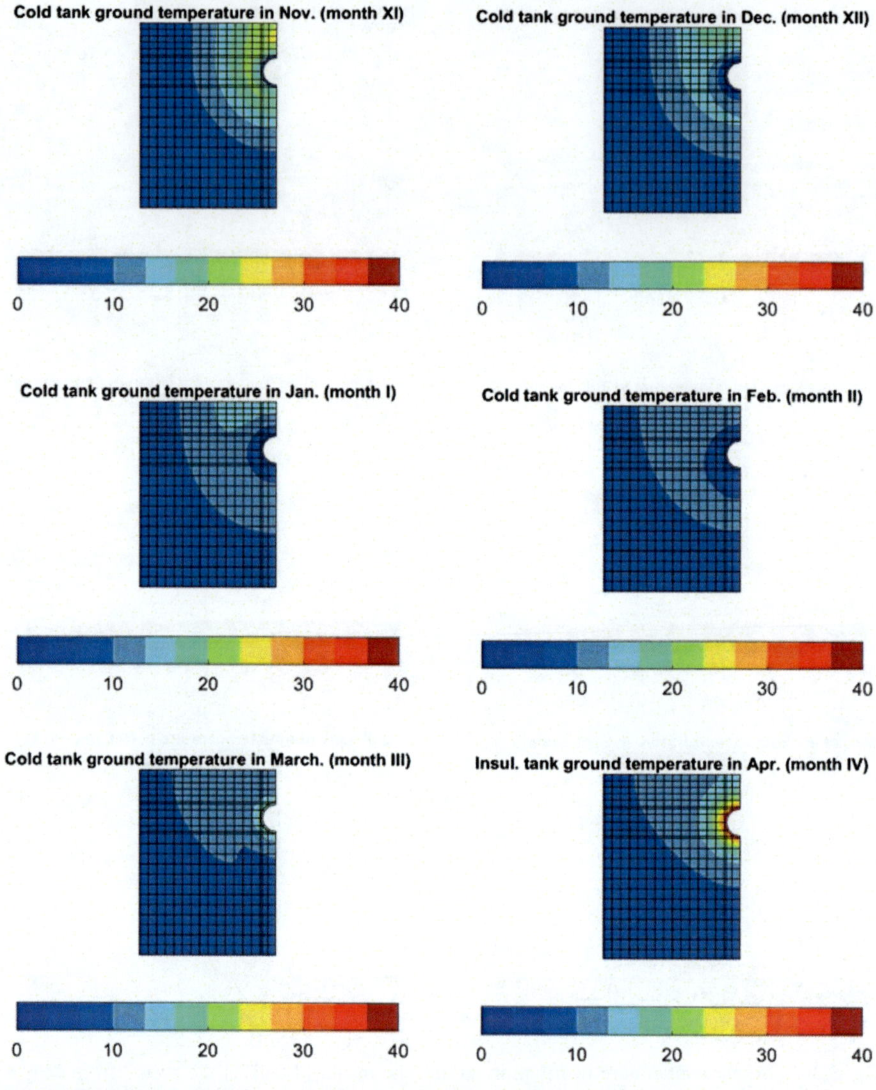


**Fig. 5.31** Ground temperature distribution around the non-insulated tank in the period from November to April (the last day of the month)—for the accumulation tank diameter $D = 2$ m and the ground initial temperature of 7 °C

heat accumulation in the ground was taken into account when calculating the average COP. The following relation was adopted for the calculations:

$$\mathrm{COP}_{year} = \frac{\int_o^{t_{year}} COPd\tau}{\int_o^{t_{year}} d\tau} \tag{5.55}$$

where the momentary COP is determined based on the chart presented in Fig. 5.12.

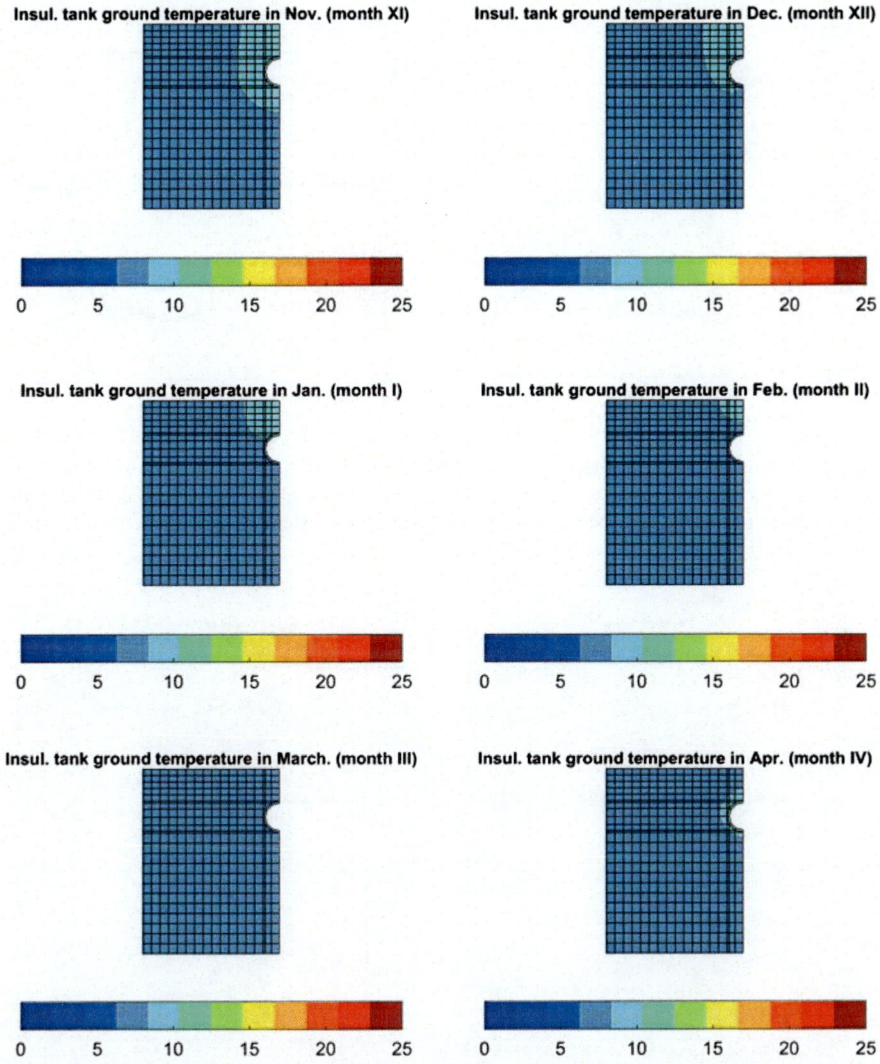


**Fig. 5.32** Ground temperature distribution around the insulated tank in the period from November to April (the last day of the month)—for the accumulation tank diameter $D = 2$ m and the ground initial temperature of 7 °C

The value of the annual average COP varies slightly with the physical properties of the ground surrounding the tank. The heat pump COP value is mainly affected by the temperature of the lower heat source.

Fig. 5.36 shows the degree of the heat demand coverage for accumulation tanks with diameter $D = 3$ m. It varies from 82% (dry soil) to 88% (moist ground). The heat pump average seasonal COP (Fig. 5.37) is higher than 4, and this value is slightly bigger compared to the system with the accumulation tank with diameter $D = 2$ m.

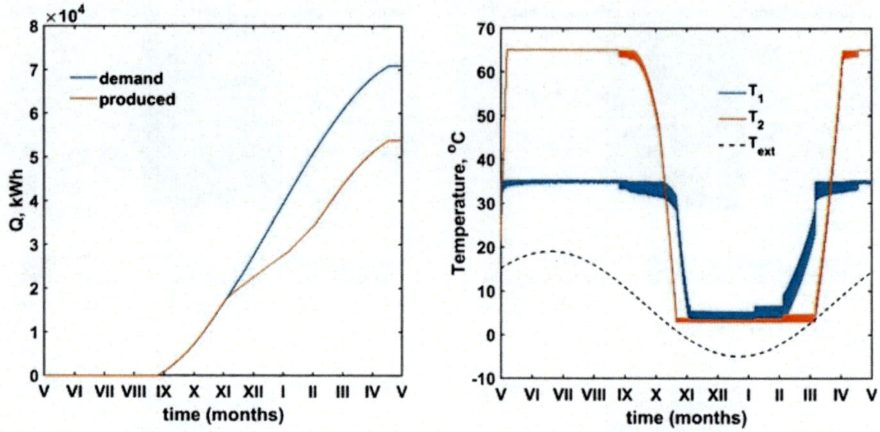


**Fig. 5.33** Building cumulated demand for thermal energy $Q_{dem}$, the amount of heat supplied to the building by the heat pump $Q_{prod}$ and the temperature distribution in the non-insulated and in the insulated tank ($T_1$ and ($T_2$, respectively in the first year of the system operation; accumulation tank diameter: $D = 2$ m, ground initial temperature: 7 °C

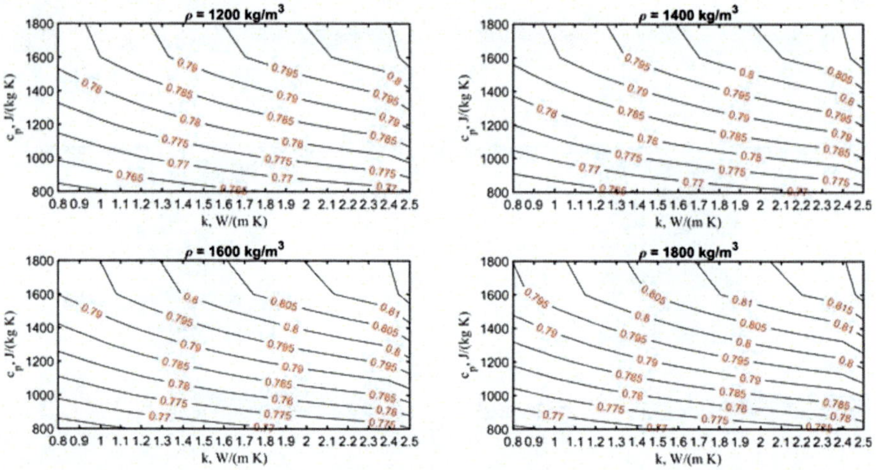


**Fig. 5.34** Building heat demand coverage degree ψ depending on the ground density, thermal conductivity and specific heat; accumulation tank diameter: $D = 2$ m; heat pump working medium: R134a

Figure 5.38 shows the values of the building heat demand coverage for diameter $D = 3$ m and the R152a cooling medium. The obtained results are only slightly different compared to refrigerant R134a. The degree of the building heat demand coverage is on average by 1–1.5% smaller compared to the R134a refrigerant.

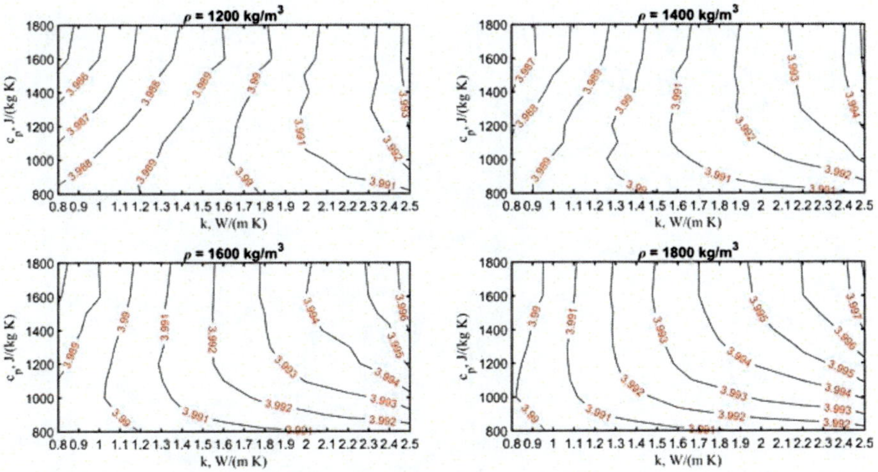


**Fig. 5.35** Value of the COP annual average ($COP_{year}$) depending on the ground density, thermal conductivity and specific heat; accumulation tank diameter: $D = 2$ m; heat pump working medium: R134a

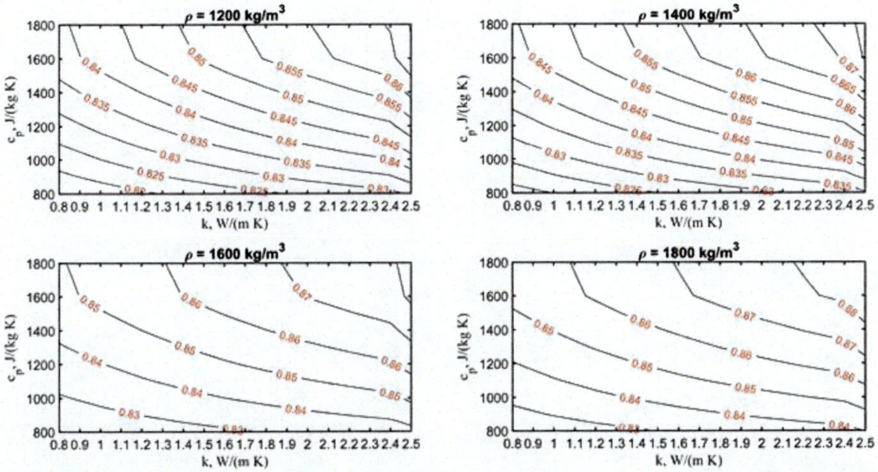


**Fig. 5.36** Building heat demand coverage degree $\psi$ depending on the ground density, thermal conductivity and specific heat; accumulation tank diameter: $D = 3$ m; heat pump working medium: R134a

By contrast, the heat pump average seasonal COP ( Fig. 5.39) is higher and takes values from 4.31 to 4.33.

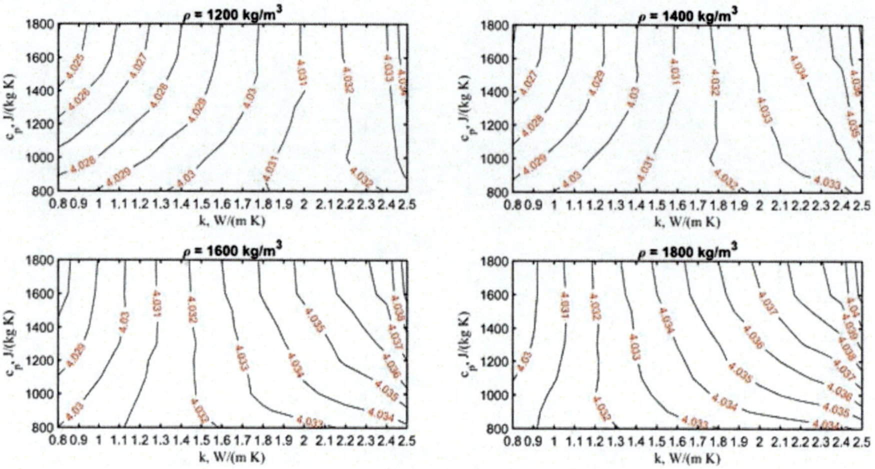


**Fig. 5.37** Value of the COP annual average ($COP_{year}$) depending on the ground density, thermal conductivity and specific heat; accumulation tank diameter: $D = 3$ m; heat pump working medium: R134a

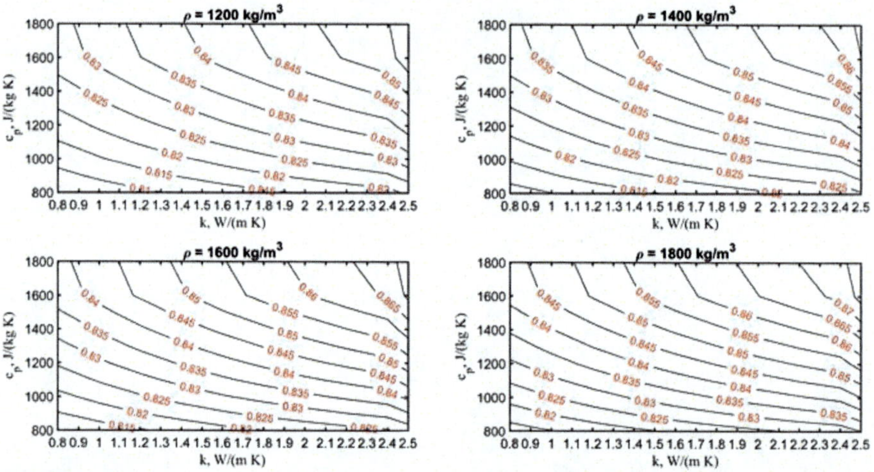


**Fig. 5.38** Building heat demand coverage degree $\psi$ depending on the ground density, thermal conductivity and specific heat; accumulation tank diameter: $D = 3$ m; heat pump working medium: R152a

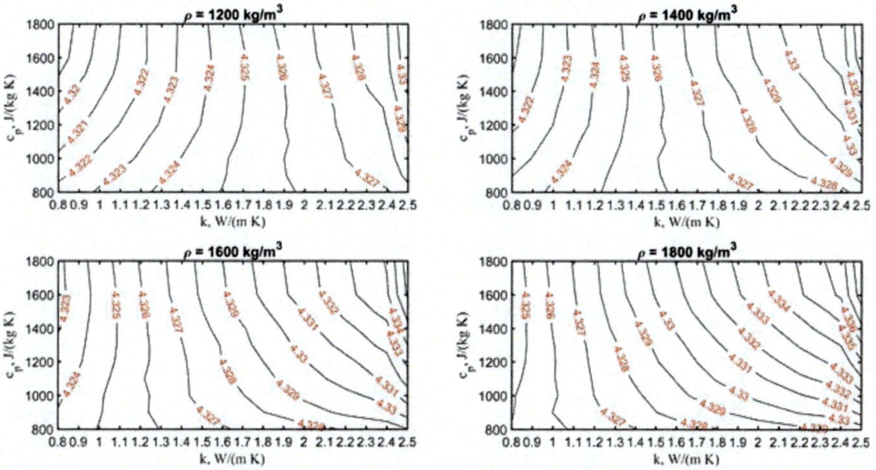


**Fig. 5.39** Value of the COP annual average ($COP_{year}$) depending on the ground density, thermal conductivity and specific heat; accumulation tank diameter: $D = 3$ m; heat pump working medium: R152a

# References


Campbell GS, Jungbauer JD, Bidlake WR, Hungerford RD (1994) Predicting the effect of temperature on soil thermal conductivity. Soil Sci 158(5):307-313

de Vries DA (1963) Thermal properties of soil chapter. In: Van Wijk WR (ed) Physics of plant environment, North Holland Pub. Co., Amsterdam, pp 210–235

Kodešová R, Vlasakova M, Fér M, Tepla D, Jakšík O, Neuberger P, Adamovsky R (2013) Thermal properties of representative soils of the czech republic. Soil Water Res 8:141–150

Kusuda T, Achenbach PR (1965) Earth Temperature and Thermal Diffusivity at Selected Stations in the United States. National Bureau of Standards; Gaithersburg, MD, USA

Rerak M (2020) PhD Thesis: A new algorithm to assist in the design of underground cable lines in view of their thermal operating conditions (in Polish)

Taler J, Duda P (2006) Solving Direct and Inverse Heat Conduction Problems, Springer, Berlin, ISBN: 978-3-540-33470-5

Usowicz B (2015) Szacowanie cieplnych właściwości gleby. Acta Agrophysica 95:352–370

# Chapter 6
# Resheat System Optimization

This chapter presents the algorithm for the RESHeat system optimization. The optimized quantity is heat extraction from the system non-insulated tank ($\phi(\tau)$) varied between 0 and 1, over a period of one year of the system operation. The optimization problem is presented below.

The RESHeat system mathematical model (Eqs. 5.17–5.30) used in the optimization process is modified as follows:

$$\rho_w c_w \frac{\pi D^2}{4} L \frac{dT_1}{d\tau} = Q_{solPV} - \phi(\tau) Q_{HP} + Q_{ground,1} \tag{6.1}$$

$$\rho_w c_w \frac{\pi D^2}{4} L \frac{dT_2}{d\tau} = Q_{solSC} - (1 - \phi(\tau)) Q_{HP} + Q_{ground,2} \tag{6.2}$$

The maximized quantity is the degree of heat demand coverage, defined as:

$$\max(\psi(\phi(\tau)) = \max\left(\frac{\int_o^{\tau_{year}} Q_{prod}(\phi(\tau))d\tau}{\int_o^{\tau_{year}} Q_b d\tau}\right) \tag{6.3}$$

where $Q_{prod}$ is calculated as:

$$\mathrm{Q}_{prod} = \int_o^{\tau_{year,1}} \phi(\tau) Q_{HP} d\tau + \int_o^{\tau_{year,2}} (1 - \phi(\tau)) Q_{HP} d\tau \tag{6.4}$$

where $\tau_{year,1}$ is the time period in the year when heat is extracted from the non-insulated tank, whereas $\tau_{year,2}$ is the time period of heat extraction from the insulated tank in the year.

The Particle Swarm Optimization Method, which is a heuristic optimization algorithm, was used to carry out the optimization. The method is gradient-free and enables iterative calculation of values of decision variables and the objective function.



P. Ocłoń, *Renewable Energy Utilization Using Underground Energy Systems*, Lecture Notes in Energy 84, https://doi.org/10.1007/978-3-030-75228-6_6

The decision variables are expressed as:

$$\phi = [\phi(\tau_1), \phi(\tau_2), \phi(\tau_3), \phi(\tau_4), \phi(\tau_5), \ldots., \phi(\tau_N)] \tag{6.5}$$

The minimized objective function is defined as follows:

$$\min(\mathrm{F}(\phi(\tau)) = \min\left(\frac{\int_o^{year} Q_b \, d\tau}{\int_o^{year} Q_{prod}(\phi(\tau))d\tau}\right) \tag{6.6}$$

## 6.1 PSO Method Algorithm

The Particle Swarm Optimization algorithm was developed in 1995 by (Kennedy and Eberhart 1995). The idea draws on mimicking the behaviour of different populations of individuals. A single individual has a limited ability to make decisions and communicate with other individuals. Individuals join together in swarms of particles or herds (typically the individuals belong to the same species) and their typical objective is the search for food or reproduction. The entire population has no central system to control them, but it has some features of intelligence in that it reacts to changes in the environment and jointly undertakes actions related thereto. Individuals (particles) change their location and thereby adapt themselves to new conditions. In this way the search for the optimum proceeds. Once a given individual has achieved a suitable position (the possibly lowest value of the objective function), it can pass on the information to all individuals or only to a part of them. Using the obtained information, the rest of the population can also achieve a position suitable for them. As a result, the whole set of particles approaches the point with the objective function best values. The block diagram of the particle swarm method algorithm is shown in Fig. 6.1 (Rerak 2020).

In the Particle Swarm Method algorithm, the velocity of each particle is calculated according to the following formula:

$$V'_{j,k,i} = wV_{j,k,i} + c_1 r_{1,j,i}\left(X_{j,pbest,i} - X_{j,k,i}\right) + c_2 r_{2,j,i}\left(X_{j,gbest,i} - X_{j,k,i}\right) \tag{6.7}$$

where

$w$ is the inertia coefficient calculated from the following equation:

$$w = w_{\max} - \frac{j(w_{\max} - w_{\min})}{j_{\max}} \tag{6.8}$$

where

$j$ number of iterations,

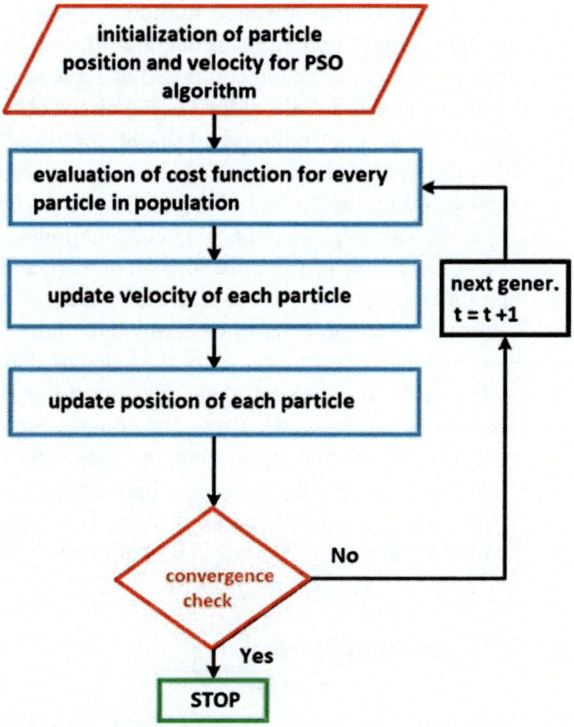


**Fig. 6.1** Block diagram of the particle swarm algorithm operation (Oclon et al. 2021)

$j_{max}$ maximum number of iterations,
$w_{max} = 0.9$,
$w_{min} = 0.4$,
$r_{1,j,i}$ and $r_{2,j,i}$ are random numbers between 0 and 1.

Symbols $c_1 = 1.5$ and $c_2 = 1.5$ in the equation are acceleration coefficients. The value of $c_1$ indicates the validity of the best local (particle) solution (*pbest*) whereas the value of $c_2$ indicates the validity of the best global solution (*gbest*). The solution is updated in each subsequent iteration in the following way:

$$X'_{j,k,i} = X_{j,k,i} + V'_{j,k,i} \tag{6.9}$$

The Particle Swarm Method has many variations inspired by various intuitive species-specific biological mechanisms. Observation of the fauna resulted in such variations of the PSO as:

(a) the bird algorithm—it takes account of the flocking behaviour of birds, which through observation and communication pass on information to each other and thus improve the process of exploring the area and finding optimal solutions,

(b) the ant algorithm—inspired by how ants look for food for their colonies. It mainly helps in finding paths in graphs. It also shows a problem solving system that is very complex but composed of basically simple but interacting units,
(c) the bee algorithm—developed based on the observation of honeybees. Honeybees live in large colonies of 20–80 thousand individuals. Their activity is coordinated by pheromones and depends on their social role/status. There are, among others, *scout* bees, i.e. individuals selected to search the area for food sources with large amounts of nectar and pollen. They determine where *recruit* bees will be "sent". The mechanism of dispatching a scout bee enables appropriate exploration of the area in the vicinity of hives and prevents wasting the strength of recruit bees by directing them immediately to the right place,
(d) the cuckoo algorithm—mimics the behaviour of some cuckoo birds which use other birds' nests to lay their eggs and raise their chicks. The most important aspect of the algorithm is that consecutive solutions (nests chosen by the cuckoos) cannot be chosen side by side. The cuckoos may, however, try to lay an egg in the same nest a second time. This does not change the fact that new solutions should generally be generated at some distance from each other, which ensures that the system will not be trapped in a local optimum.

## 6.2 Calculation Results

The calculations are performed in the MATLAB software package. Figure 6.2 shows the optimum value of objective function $F(\phi)$ and the degree of coverage of the demand for heat for consecutive iterations of the particle swarm method algorithm. The calculations are carried out for the accumulation tank diameter of $D = 3$ m. The following values of the ground thermal properties are assumed: $k = 1$ W/(m K), $\rho =$

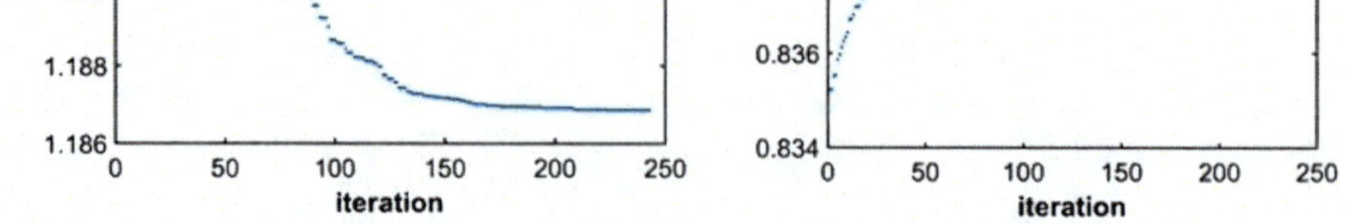


**Fig. 6.2** Optimum value of objective function F(ϕ) and the degree of coverage of heat demand $\Psi$ in consecutive iterations for $N = 20$ particles; accumulation tank diameter $D = 3$ m

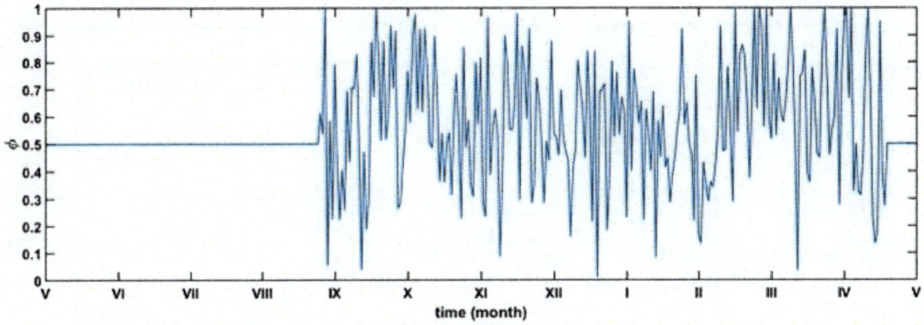


**Fig. 6.3** Curve illustrating changes in the value of variable $\phi(\tau)$ during the heating season obtained for the optimal solution; tank diameter equal to $D = 3$ m

1600 kg/m$^3$, $c_p = 1400$ J/(kg K). The sampling period for the performed calculations totals 1 h. The number of particles is $N = 20$.

For the values of $\phi(\tau)$ shown in Fig. 6.3, the following histories are obtained of the temperature of water in the accumulation tanks and of the produced and required amount of heat for the building (Figs. 6.4 and 6.5, respectively).

Figure 6.6 shows the values of objective function $F(\phi)$ and the degree of coverage of the demand for heat for consecutive iterations of the particle swarm method algorithm. The calculations are carried out for the accumulation tank diameter of $D =$ 2 m.

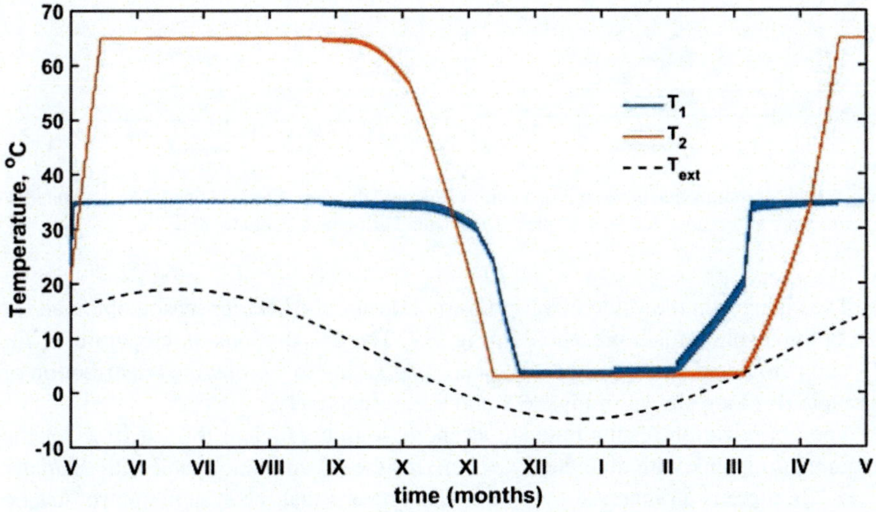


**Fig. 6.4** History of the temperature of water in the non-insulated tank $T_1$, in the insulated tank $T_2$ and external temperature $T_{\text{ext}}$ for the $\phi(\tau)$ values obtained from the optimization; tank diameter equal to$D = 3$ m

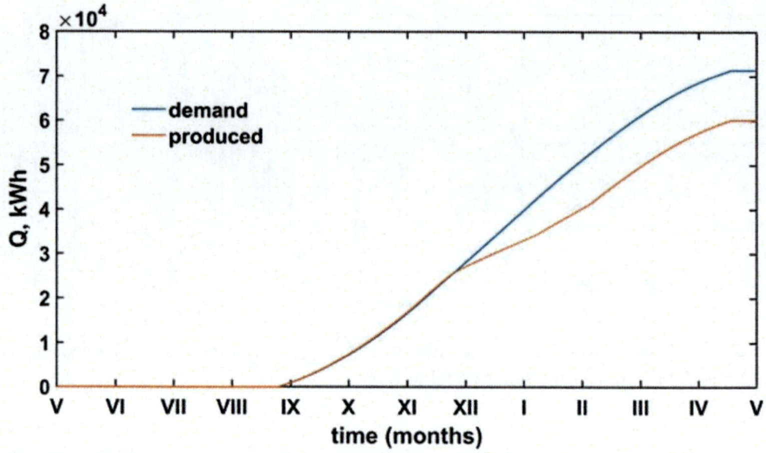


**Fig. 6.5** Building cumted demand for heat $Q_{dem}$. and the amount of heat $Q_{prod}$, supplied to the building by the heat pump in the first year of the system operation for $D = 3$ m

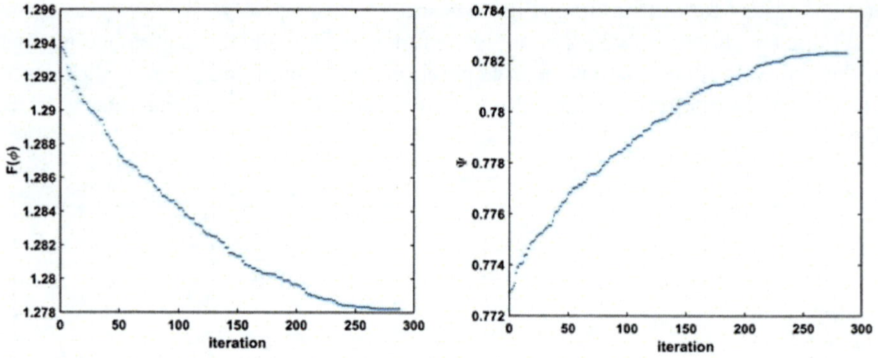


**Fig. 6.6** Optimum value of objective function $F(\phi)$ and the degree of coverage of heat demand $\Psi$ in consecutive iterations for $N = 20$ particles; accumulation tank diameter $D = 2$ m

The changes in the value of variable $\phi(\tau)$ during the heating season obtained for the optimal solution are presented in Fig. 6.7. The distributions of temperatures $T_1$, $T_2$ and cumulated heat $Q_{prod}$ and $Q_{dem}$ corresponding to the obtained distribution of variable $\phi(\tau)$ are shown in Figs. 6.8 and 6.9, respectively.

The optimization results indicate an increase in coefficient $\psi$ by 1.01% for the accumulation tank with diameter $D = 3$ m; for the accumulation tank with diameter $D = 2$ m there is an increase by 1.42%. This means that the system control has no significant impact on the degree of the RESHeat system coverage of the building demand for heat. However, it slightly increases its value.

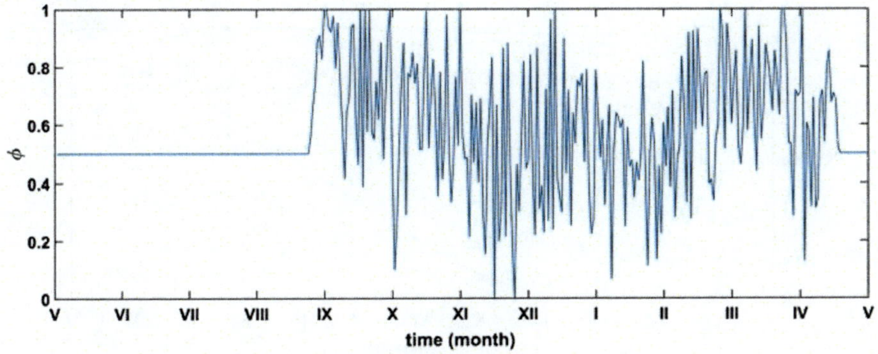


**Fig. 6.7** Curve illustrating changes in the value of variable $\phi(\tau)$ during the heating season obtained for the optimal solution; accumulation tank diameter $D = 2$ m

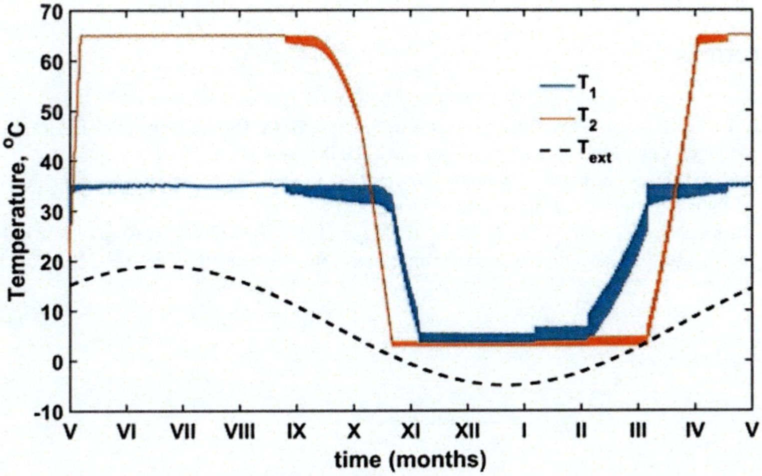


**Fig. 6.8** History of the temperature of water in the non-insulated tank $T_1$, in the insulated tank $T_2$ and external temperature $T_{ext}$ for the $\phi(\tau)$ values obtained from the optimization; tank diameter $D = 2$ m

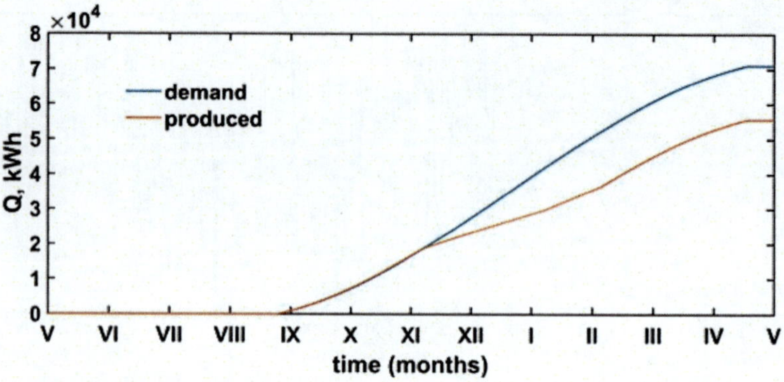


**Fig. 6.9** Building cumulated demand for heat $Q_{\mathrm{dem}}$ and the amount of heat $Q_{\mathrm{prod}}$, supplied to the building by the heat pump in the first year of the system operation for $D = 3$ m and the values of $\phi(\tau)$ obtained from the optimization

## References

Kennedy J, Eberhart R (1995) Particle swarm optimization. In: Proceedings of IEEE international conference on neural networks, vol IV, pp 1942–1948

Rerak M (2020) PhD Thesis: A new algorithm to assist in the design of underground cable lines in view of their thermal operating conditions (in Polish)

Ocłoń P, Łopata S, Stelmach T, Li M, Zhang JF, Mzad H, Tao WQ (2021) Design optimization of a high-temperature fin-and-tube heat exchanger manifold—a case study. ENERGY 215: 119059

# Chapter 7
# Modelling Heat Transfer in the PV Panel Cooling System

Photovoltaic thermal (PVT) cells are one of the key elements of the RESHeat system. The amount of heat accumulated in the ground is determined by the efficiency of solar-to-thermal energy conversion, and the amount of produced electricity depends on the efficiency of the cooling system. Figure 7.1 shows how the cooling system segments are installed on the photovoltaic panels. They are mounted by pressing two beams with spring elements against the panels. Using this installation method, the ELFRAN company avoids additional thermal stresses. The ELFRAN cooling system does not in any way violate the guarantee terms and conditions specified by the PV panel manufacturer. It is therefore universal for each type of PV panels. Thermal stresses are eliminated by flexible connectors (Fig. 7.2) in the temperature range from −30 °C to 70 °C (Kozak-Jagieła 2020).

A laboratory stand with PV panels was mounted on the roof of one of the buildings of the Faculty of Mechanical Engineering of the Cracow University of Technology. It was made under the HySOL project (a project for bilateral Polish-German collaboration financed by the National Centre for Research and Development—NCRD). The PV panels used in the stand are the ECO DELTA 280P PV panels with the power capacity of up to 280 W. The panels are mounted both on the sun-tracker (rotating panels equipped with the sun-tracking mechanism) and in the stationary system. The surface of the panels is divided in such a manner that one half of the surface is cooled whereas the other half is not. Figure 7.3a, b show the prototype of the PV cooling system mounted on the sun-tracker and on the stationary panels, respectively.

The test stand is metered using the following devices:

1. weather station (measurement of atmospheric pressure, temperature, air humidity, wind speed),
2. pyranometer (measurement of solar radiation intensity) for both stationary panels and those mounted on the tracker,
3. current and voltage transducers (measurement of the rms voltage value),
4. jacket thermocouples type K with the jacket diameter of 0.5 mm and the length of 1,000 mm (temperature measurement in the range from −40 to 1,000 °C).



P. Ocłoń, *Renewable Energy Utilization Using Underground Energy Systems*, Lecture Notes in Energy 84, https://doi.org/10.1007/978-3-030-75228-6_7

Fig. 7.1 Cooling system installed for stationary PV panels (Kozak-Jagieła 2020)

Fig. 7.2 Cooling system installed for PV panels with the sun-tracker (Kozak-Jagieła 2020)

Figure 7.4 presents the arrangement of the temperature measuring points of the panels and of the cooling system.

## 7.1 Numerical Modelling of the Temperature Distribution of PVT Panels

This chapter presents the mathematical model of the heat exchange for PV panels with a cooling system. The heat conduction model used herein is three-dimensional and enables analysis of transient-state phenomena. A one-dimensional thermal energy transport equation is used for the medium cooling the PVT panels.

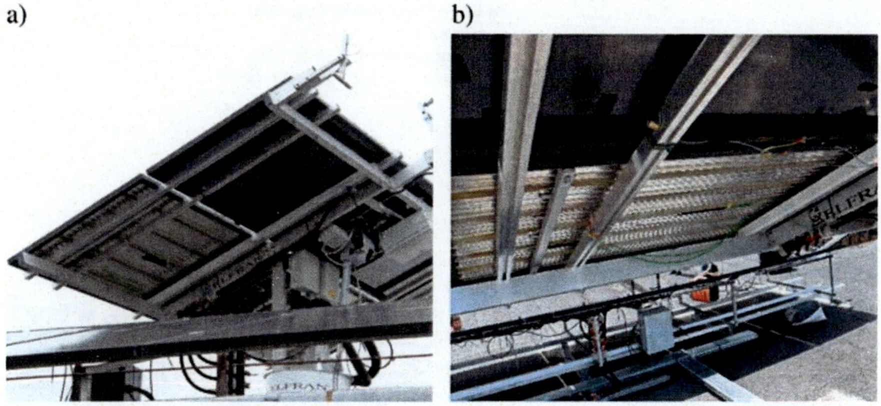


**Fig. 7.3** Prototype of the PV panels cooling system: **a** mounted on the sun-tracker, **b** mounted on the stationary panels (Kozak-Jagieła 2020)

### *7.1.1 Model of Heat Exchange in PVT Panels*

In the Cartesian system of coordinates transient heat conduction for the PV panel layers is described by Eqs. (7.1)–(7.5). Equation (7.6) describes the heat conduction process in the radiator tubes and is expressed in the system of cylindrical coordinates (Oclon et al, 2020):

- PV panel area:

$$\rho_{PV} c_{PV} \frac{\partial T}{\partial \tau} = \frac{\partial}{\partial x}\left(k_{PV}\frac{\partial T}{\partial x}\right) + \frac{\partial}{\partial y}\left(k_{PV}\frac{\partial T}{\partial y}\right) + \frac{\partial}{\partial z}\left(k_{PV}\frac{\partial T}{\partial z}\right) + \dot{q}_v \tag{7.1}$$

where the volume heat source is defined as:

$$\dot{q}_v = \frac{G \cdot (1 - \eta_{PV}) \cdot \eta_{abs}}{\Delta y_{PV}} \tag{7.2}$$

where $\Delta y_{PV}$ is the thickness of the PV panel material layer.

- Glass layer area:

$$\rho_{sz} c_{sz} \frac{\partial T}{\partial \tau} = \frac{\partial}{\partial x}\left(k_{sz}\frac{\partial T}{\partial x}\right) + \frac{\partial}{\partial y}\left(k_{sz}\frac{\partial T}{\partial y}\right) + \frac{\partial}{\partial z}\left(k_{sz}\frac{\partial T}{\partial z}\right) \tag{7.3}$$

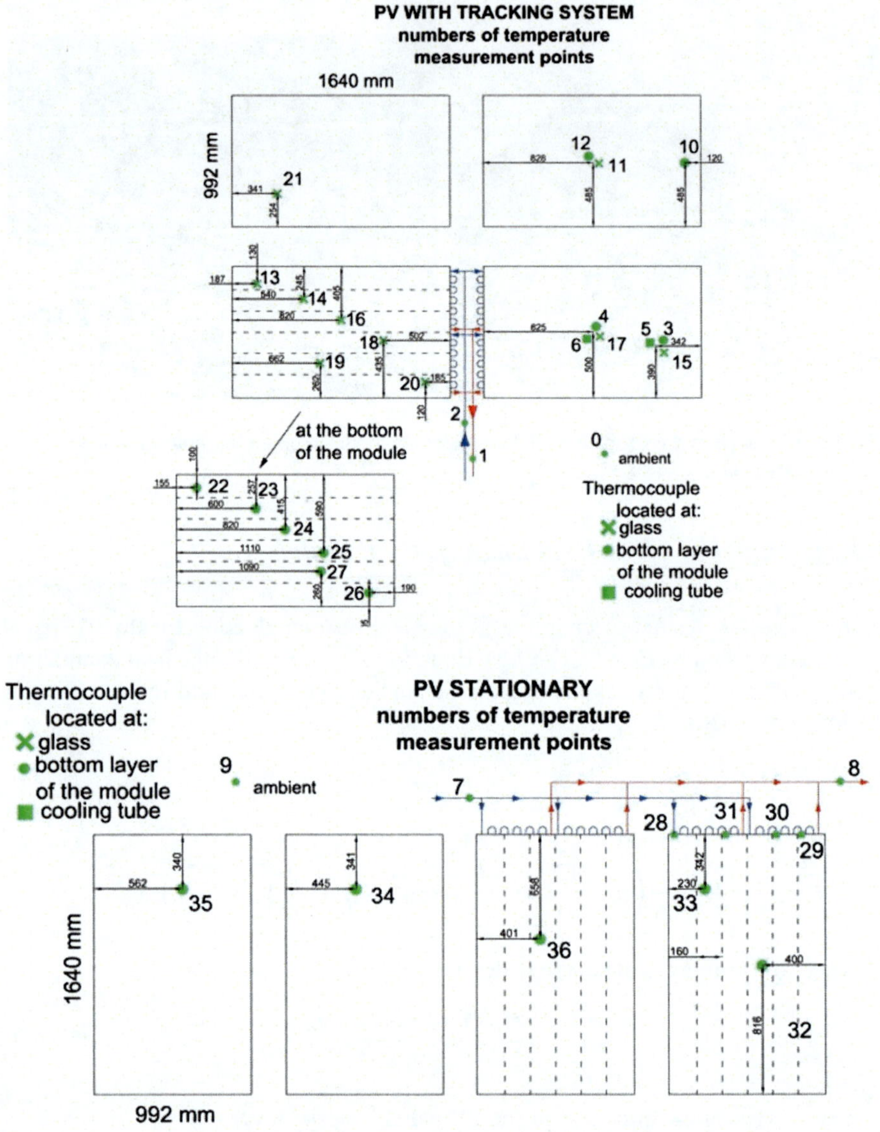


Fig. 7.4 Distribution of the temperature measuring points on the stationary panels and on the sun-tracker, both cooled and non-cooled (Kozak-Jagieła 2020)

- EVA film area:

$$\rho_f c_f \frac{\partial \mathrm{T}}{\partial \tau} = \frac{\partial}{\partial \mathrm{x}}\left(k_f \frac{\partial T}{\partial x}\right) + \frac{\partial}{\partial y}\left(k_f \frac{\partial T}{\partial y}\right) + \frac{\partial}{\partial z}\left(k_f \frac{\partial T}{\partial z}\right) \tag{7.4}$$

- Radiator area:

$$\rho_{al}c_{al}\frac{\partial \mathrm{T}}{\partial \tau} = \frac{\partial}{\partial \mathrm{x}}\left(k_{al}\frac{\partial T}{\partial x}\right) + \frac{\partial}{\partial y}\left(k_{al}\frac{\partial T}{\partial y}\right) + \frac{\partial}{\partial z}\left(k_{al}\frac{\partial T}{\partial z}\right) \tag{7.5}$$

- Area of the cooling system aluminium tubes:

$$\rho_{al}c_{al}\frac{\partial \mathrm{T}}{\partial \tau} = \frac{1}{\mathrm{r}}\frac{\partial}{\partial \mathrm{r}}\left(rk_{al}\frac{\partial T}{\partial r}\right) + \frac{1}{r^2}\frac{\partial}{\partial \theta}\left(k_{al}\frac{\partial T}{\partial \theta}\right) + \frac{\partial}{\partial z}\left(k_{al}\frac{\partial T}{\partial z}\right) \tag{7.6}$$

where

| | |
|---|---|
| $G$ | direct solar radiation intensity, W/m$^2$ |
| $\eta_{PV}$ | electrical efficiency of the cells, $\eta_{PV} = 18\%$ (as per the technical data sheet), |
| $\eta_{abs}$ | heat radiation absorption efficiency (assumed) $\eta_{abs} = 97\%$, |
| $k$ | heat conductivity coefficient, W/(m K), |
| $\rho$ | density, kg/m$^3$ |
| $c$ | specific heat, J/(kg K) |
| $T$ | temperature, K |
| $t$ | time, s |
| $x, y, z$ | Cartesian coordinates, m |
| $r, \theta$ | cylindrical coordinates, m; rad |

The following boundary conditions are used in the model:

- Convective heat flow between the panel wall and air:

  For the aluminium radiator external surface :

$$-k_{al}\frac{\partial T}{\partial y} = h_e(T - T_{ext}), \tag{7.7}$$

  for the glass external surface:

$$-k_{sz}\frac{\partial T}{\partial y} = h_e(T - T_{ext}), \tag{7.8}$$

where:

| | |
|---|---|
| $h_e$ | heat transfer coefficient on the air side (heat transfer from external air to the panel/radiator wall), |
| $T_{ext}$ | external (ambient) temperature, K. |

Surface film conductance $h_e$ is calculated using formula (7.9) (Kozak-Jagieła 2020) :

$$h_e = 10.45 - u_e + \sqrt{u_e}, \tag{7.9}$$

where

$u_e$ wind speed, m/s

- The convective heat flow between external air and the tube outer surface is expressed using formula (7.10):

$$-k_{al}\left.\frac{\partial T}{\partial r}\right|_{r=r_{out}} = h_e(T - T_{ext}) \tag{7.10}$$

The heat flux on the fluid–solid body interface is determined by the convective boundary condition described by formula (7.11) below.

- Condition of the convective heat transfer in the tube wall-cooling fluid area:

$$-k_{al}\left.\frac{\partial T}{\partial r}\right|_{r=r_i} = h_{in}(T - T_{in}) \tag{7.11}$$

The convective heat transfer coefficient $h_{in}$ of the of coolant with temperature $T_{in}$ flowing in the tubes is calculated using the Gnielinski correlation (Taler and Ocłoń 2014). The correlation is used to find the Nusselt number (Nu). It is characterized by good accuracy in the turbulent and transient flow range. The Nusselt number in the heat exchange determines the ratio between the rates of the convective and the conductive heat transfer. The higher the Nu number, the faster the effect of the heat transfer due to convection.

The Nusselt number can be described using correlation (7.12) (Kozak-Jagieła 2020):

$$\mathrm{Nu} = \frac{h_{in}d}{k} = \frac{\frac{\xi}{8}(\mathrm{Re} - 1000)\mathrm{Pr}}{1 + 12.7\sqrt{\frac{\xi}{8}}\left(\mathrm{Pr}^{\frac{2}{3}} - 1\right)}$$
$$3 \cdot 10^3 \leq \mathrm{Re} \leq 5 \cdot 10^6,\ 0.5 \leq \mathrm{Pr} \leq 200, \tag{7.12}$$

where:

$\xi$ friction factor for plain (smooth) tubes

The Reynolds number (Re) in formula (7.12) describes the mechanical phenomena occurring due to internal friction. It defines the ratio between the forces of inertia and internal friction occurring in the analysed conditions. The following types of flow are assumed in engineering calculations depending on the Re number value:

- laminar flow for Re < 2100
- transient flow for 2100 < Re < 3000

- turbulent flow for Re > 3000.

The Reynolds number can be described using the following formula:

$$Re = \frac{u_m \cdot d_w}{\nu} \tag{7.13}$$

where: $u_m$—fluid flow mean velocity in the tube cross-section; $d_w$—inner diameter; $\nu$—kinematic viscosity.

The Prandtl number (Pr) defines the ratio between the fluid viscosity and the fluid thermal conductivity. It is expressed as follows: (Taler and Ocłoń 2014)

$$Pr = \frac{c \cdot \mu}{k} \tag{7.14}$$

where: $c$—specific heat; $\mu$—dynamic viscosity coefficient; $k$—heat conductivity coefficient.

Knowing the coefficient of the heat transfer from the fluid to the tube inner wall $h_{in,}$ the coolant temperature $T_{in}$ and the wall temperature $T$, it is possible to determine the heat flux on the fluid–solid body interface using Eq. (7.15).

The conditions on the interface of the PVT panel layers are described below.

- PV panel-glass interface:

$$-k_{sz}\left.\frac{\partial T}{\partial y}\right|_{y=y_{sz}} = -k_{PV}\left.\frac{\partial T}{\partial y}\right|_{y=y_{sz}} \tag{7.15}$$

- PV panel-EVA film interface:

$$-k_{PV}\left.\frac{\partial T}{\partial y}\right|_{y=y_{PV}} = -k_{f}\left.\frac{\partial T}{\partial y}\right|_{y=y_{PV}} \tag{7.16}$$

- EVA film-radiator interface:

$$-k_{f}\left.\frac{\partial T}{\partial y}\right|_{y=y_{f}} = -k_{al}\left.\frac{\partial T}{\partial y}\right|_{y=y_{f}} \tag{7.17}$$

- adhesive layer-radiator interface:

$$-k_{al}\left.\frac{\partial T}{\partial y}\right|_{y=y_{al}}=-k_{glue}\left.\frac{\partial T}{\partial y}\right|_{y=y_{al}} \tag{7.18}$$

- adhesive layer-tube outer surface interface:

$$-k_{glue}\left.\frac{\partial T}{\partial r}\right|_{r=r_o}=-k_{al}\left.\frac{\partial T}{\partial r}\right|_{r=r_0} \tag{7.19}$$

The energy conservation equation for the cooling medium is expressed using the following formula :

$$\rho_{in}c_{in}\frac{\partial T_{in}}{\partial \tau}+\rho_{in}c_{in}u_{in}\frac{\partial T_{in}}{\partial z}=k_{in}\frac{\partial^2 T_{in}}{\partial z^2}-\frac{2h_{in}}{r_i}(T_{in}-T) \tag{7.20}$$

The following boundary condition is adopted for the cooling medium inlet:

$$z=0: T_{in}(\tau, z=0)=T_{inlet} \tag{7.21}$$

where: $T_{inlet}$—coolant inlet temperature, with the initial condition:

$$T_{in}(\tau=0, z)=T_0 \tag{7.22}$$

$T_0$—initial temperature at the inlet read based on measurements.
Time step $\Delta\tau$ is found from the Courant-Friedrichs-Lewy condition:

$$\Delta\tau<\frac{\Delta z}{u_{in}} \tag{7.23}$$

$\Delta z$ numerical mesh spatial increment in the direction of axis $z$.

The cooling medium velocity $u_{in}$ is established using the following formula:

$$u_{in}=\frac{m_{in}}{\pi\cdot r_{in}^2\cdot\rho}=\frac{Q_v}{\pi\cdot r_{in}^2} \tag{7.24}$$

where:

$m_{in}$ cooling medium mass flow rate, kg/s.
$Q_v$ cooling medium volumetric flow rate, $m^3/s$.

### 7.1.2 PV Panel Energy Balance Equation—The Cartesian System

Figure 7.5 shows a rectangular prism with the dimensions: $\Delta x$, $\Delta y$, and $\Delta z$, and with eight nodes. Control volumes are created around each node and energy balance equations are written for them. Figure 7.6 presents the discretization method of the three-dimensional heat conduction equation.

The balance equations for individual nodes are presented below:

Energy balance equations for the glass area (Node 1):

$$Q_{2-1} = k_{sz} \cdot \frac{\Delta y_{sz}}{2} \cdot \frac{\Delta z}{2} \cdot \frac{(T_2 - T_1)}{\Delta x}$$

$$Q_{4-1} = k_{sz} \cdot \frac{\Delta x}{2} \cdot \frac{\Delta z}{2} \cdot \frac{(T_4 - T_1)}{\Delta y_{sz}}$$

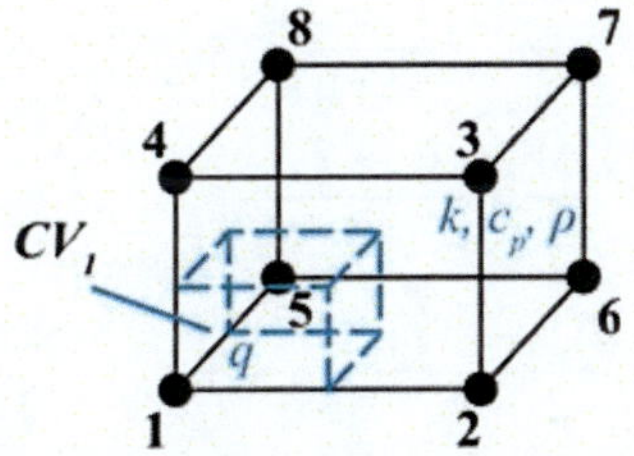


Fig. 7.5 Three-dimensional diagram of the PV module with node numbering

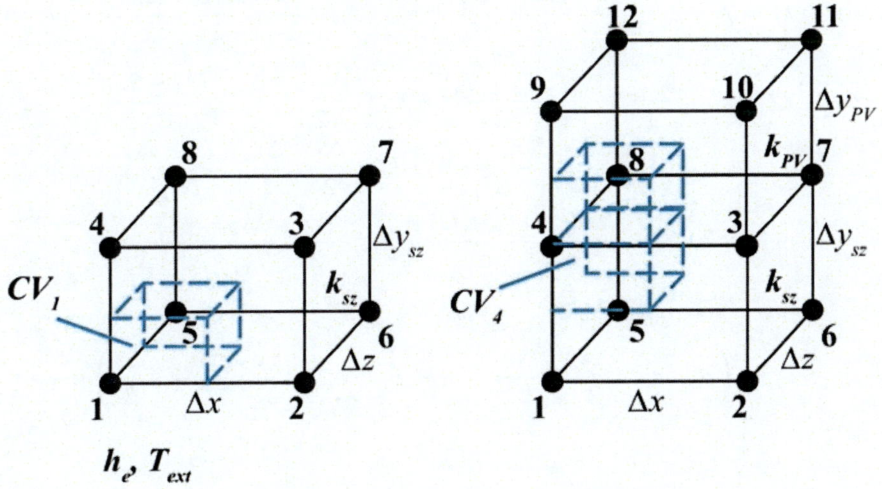


Fig. 7.6 Discretization—the Cartesian system: **a** glass, **b** glass/PV panel

$$Q_{5-1} = k_{sz} \cdot \frac{\Delta y_{sz}}{2} \cdot \frac{\Delta x}{2} \cdot \frac{(T_5 - T_1)}{\Delta z} \tag{7.25}$$

$$Q_{ext-1} = h_e \cdot \frac{\Delta x}{2} \cdot \frac{\Delta z}{2} \cdot (T_{ext} - T_1)$$

$$M = \rho \cdot c_s \cdot \frac{\Delta x}{2} \cdot \frac{\Delta y_{sz}}{2} \cdot \frac{\Delta z}{2} \tag{7.26}$$

$$\begin{aligned}\sum Q = k_{sz} \cdot \frac{\Delta y_{sz}}{2} \cdot \frac{\Delta z}{2} \cdot \frac{(T_2 - T_1)}{\Delta x} + k_{sz} \cdot \frac{\Delta x}{2} \cdot \frac{\Delta z}{2} \cdot \frac{(T_4 - T_1)}{\Delta y_{sz}} + k_{sz} \cdot \frac{\Delta y_{sz}}{2} \cdot \frac{\Delta x}{2} \cdot \frac{(T_5 - T_1)}{\Delta z} \\ + h_e \cdot \frac{\Delta x}{2} \cdot \frac{\Delta z}{2} \cdot (T_{ext} - T_1) = \rho_s \cdot c_s \cdot \frac{\Delta x}{2} \cdot \frac{\Delta y_{sz}}{2} \cdot \frac{\Delta z}{2} \cdot \frac{dT_1}{d\tau}\end{aligned} \tag{7.27}$$

- The energy balance equations in the glass/PV panel system (Node 4) can be expressed as:

$$\begin{aligned}Q_{3-4} &= k_S \cdot \frac{\Delta y_s}{2} \cdot \frac{\Delta z}{2} \cdot \frac{(T_3 - T_4)}{\Delta x} + k_{pv} \cdot \frac{\Delta y_{pv}}{2} \cdot \frac{\Delta z}{2} \cdot \frac{(T_3 - T_4)}{\Delta x} \\ Q_{9-4} &= k_{pv} \cdot \frac{\Delta x}{2} \cdot \frac{\Delta z}{2} \cdot \frac{(T_9 - T_4)}{\Delta y_{pv}} \\ Q_{1-4} &= k_{sz} \cdot \frac{\Delta x}{2} \cdot \frac{\Delta z}{2} \cdot \frac{(T_1 - T_4)}{\Delta y_{sz}}\end{aligned} \tag{7.28}$$

$$\begin{aligned}Q_{4-8} &= k_{sz} \cdot \frac{\Delta x}{2} \cdot \frac{\Delta y_{sz}}{2} \cdot \frac{(T_4 - T_8)}{\Delta z} + k_{pv} \cdot \frac{\Delta x}{2} \cdot \frac{\Delta y_{pv}}{2} \cdot \frac{(T_4 - T_8)}{\Delta z} \\ Q_{pv} &= (1 - \eta_{PV}) \cdot G \cdot \frac{\Delta x}{2} \cdot \frac{\Delta z}{4} \eta_{odb}\end{aligned} \tag{7.29}$$

$$M = \rho_{sz} \cdot c_{sz} \cdot \frac{\Delta x}{2} \cdot \frac{\Delta y_{sz}}{2} \cdot \frac{\Delta z}{2} + \rho_{pv} \cdot c_{pv} \cdot \frac{\Delta x}{2} \cdot \frac{\Delta y_{pv}}{2} \cdot \frac{\Delta z}{2} \tag{7.30}$$

$$\begin{aligned}\sum Q = k_{sz} \cdot \frac{\Delta y_{sz}}{2} \cdot \frac{\Delta z}{2} \cdot \frac{(T_3 - T_4)}{\Delta x} + k_{pv} \cdot \frac{\Delta y_{pv}}{2} \cdot \frac{\Delta z}{2} \cdot \frac{(T_3 - T_4)}{\Delta x} \\ + k_{pv} \cdot \frac{\Delta x}{2} \cdot \frac{\Delta z}{2} \cdot \frac{(T_9 - T_4)}{\Delta y_{pv}} + k_{sz} \cdot \frac{\Delta x}{2} \cdot \frac{\Delta z}{2} \cdot \frac{(T_1 - T_4)}{\Delta y_{sz}} \\ + k_S \cdot \frac{\Delta x}{2} \cdot \frac{\Delta y_{sz}}{2} \cdot \frac{(T_4 - T_8)}{\Delta z} + k_{pv} \cdot \frac{\Delta x}{2} \cdot \frac{\Delta y_{pv}}{2} \cdot \frac{(T_4 - T_8)}{\Delta z} \\ + (1 - \eta_{PV}) G \frac{\Delta x}{2} \cdot \frac{\Delta z}{4} \eta_{odb} \\ = \left( \rho_{sz} \cdot c_{sz} \cdot \frac{\Delta x}{2} \cdot \frac{\Delta y_{sz}}{2} \cdot \frac{\Delta z}{2} + \rho_{pv} \cdot c_{pv} \cdot \frac{\Delta x}{2} \cdot \frac{\Delta y_{pv}}{2} \cdot \frac{\Delta z}{2} \right) \cdot \frac{dT_4}{d\tau}\end{aligned} \tag{7.31}$$

### 7.1.3 PV Panel Energy Balance Equation—The Cylindrical System

The proposed PV panel cooling system consists of aluminium tubes with the working medium in the form of a 50/50 glycol–water solution. The energy balance equation for the aluminium tubes is described in the cylindrical system of coordinates. The diagram of the cooling system using aluminium tubes is presented in Fig. 7.7 together with the node numbering.

In the area of the radiator tubes the equations are written in the cylindrical system of coordinates using formulae (7.32–7.36):

$$Q_{4-1} = k_{al} \cdot \left(r_i + \frac{\Delta r}{2}\right) \cdot \frac{\Delta\varphi}{2} \cdot \frac{\Delta z}{2} \cdot \frac{(T_4 - T_1)}{\Delta r} \quad Q_{5-1} = k_{al} \cdot \Delta\varphi \cdot r_i \cdot \Delta r \cdot \frac{(T_5 - T_1)}{4\Delta z} \tag{7.32}$$

$$Q_{2-1} = k_{al} \cdot \left(\frac{\Delta r}{2}\right) \cdot \frac{\Delta z}{2} \cdot \frac{(T_2 - T_1)}{r_i \cdot \Delta\varphi} \tag{7.33}$$

$$Q_h = h_{in} \cdot r_i \cdot \frac{\Delta\varphi}{2} \cdot (T_{in} - T_1) \tag{7.34}$$

$$M = \left(\left(r_i + \frac{\Delta r}{2}\right)^2 - (r_i)^2\right) \cdot \frac{\Delta\varphi}{4} \cdot \rho_{al} \cdot c_{al} \tag{7.35}$$

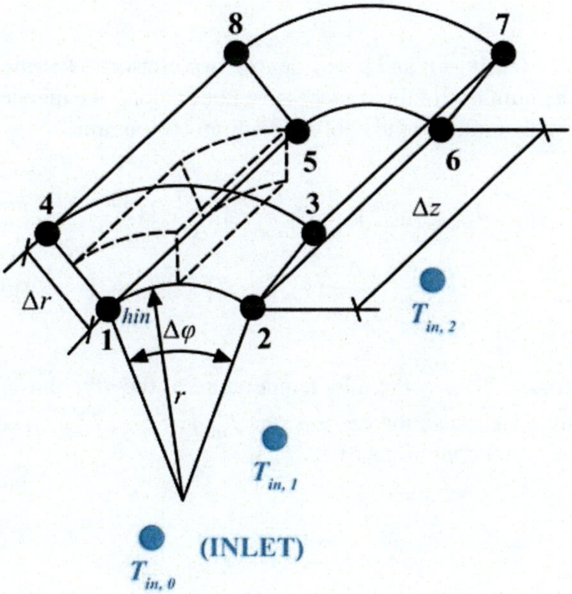


**Fig. 7.7** Three-dimensional diagram of the PV panel cooling system with node numbering (Ocłoń et al. 2020)

$$\sum Q = k_{al} \cdot \left(r + \frac{\Delta r}{2}\right)\frac{\Delta\varphi}{2} \cdot \frac{\Delta z}{2} \cdot \frac{(T_4 - T_1)}{\Delta r} + k_{al} \cdot \frac{\Delta r}{2} \cdot \frac{\Delta z}{2} \cdot \frac{(T_2 - T_1)}{r_i \cdot \Delta\varphi}$$
$$+ k_{al} \cdot \Delta\varphi \cdot r_i \cdot \Delta r \cdot \frac{(T_5 - T_1)}{4\Delta z} + h_{in} \cdot r_i \cdot \frac{\Delta\varphi}{2} \cdot (T_{in} - T_1) = \left(\left(r_i + \frac{\Delta r}{2}\right)^2 - (r_i)^2\right) \cdot \frac{\Delta\varphi}{4} \cdot \rho_{al} \cdot c_{al} \cdot \frac{dT_1}{d\tau} \tag{7.36}$$

## 7.1.4 *Discretization of the Coolant Energy Equation—Finite Difference Method*

The finite difference method consists in replacing derivatives in the equations with appropriate difference quotients. In this way it is possible to analyse various boundary conditions. However, it should be remembered that the shape of the body should be isometric. Due to its universal applicability, the method makes it possible to find the solution in areas with complex shapes.

The discretization of the time derivative in the energy conservation Eq. (7.37) is performed using the explicit method, whereas the spatial derivative discretization is carried out using the backward difference quotient. The central difference quotient is used to discretize the second-order terms.

$$\rho_{in}c_{in}\frac{\left(T^{n}_{in,1} - T^{n-1}_{in,1}\right)}{\Delta t} + \rho_{in}c_{in}u_{in}\frac{\left(T^{n-1}_{in,1} - T^{n-1}_{in,0}\right)}{\Delta z}$$
$$= k_{in}\frac{(T^{n-1}_{in,2} - 2T^{n-1}_{in,1} + (T^{n-1}_{in,0})}{\Delta z^2} - \frac{2h_{in}}{r_i}\left(T^{n-1}_{in,1} - T^{n-1}_{1}\right) \tag{7.37}$$

Markings $n$ and $n - 1$ denote two consecutive time steps, whereas 0, 1 and 2 are the numbers of three successive nodes along the tube length. Rearranging Eq. (7.37), the formula for the cooling medium temperature $T^{n}_{in,1}$ is obtained:

$$T^{n}_{in,1} = T^{n-1}_{in,1} + \left[\frac{k_{in}}{\rho_{in}c_{in}}\left(\frac{\Delta\tau}{\Delta z^2}\right)\left(T^{n-1}_{in,2} - 2T^{n-1}_{in,1} + T^{n-1}_{in,0}\right)\right.$$
$$\left. - \frac{2h_{in}\Delta\tau}{r_i\rho_{in}c_{in}}\left(T^{n-1}_{in,1} - T^{n-1}_{1}\right) - u_{in}\left(\frac{\Delta\tau}{\Delta z}\right)\left(T^{n-1}_{in,1} - T^{n-1}_{in,0}\right)\right] \tag{7.38}$$

where: $T^{n-1}_{in,0}$ is the inlet temperature of the glycol–water solution from the previous time step; also the expressions $T^{n-1}_{in,0}$, $T^{n-1}_{in,1}$, $T^{n-1}_{in,2}$ relate to the previous iteration or to initial conditions ( Fig. 7.7).

## 7.2 Analysis of the Cooling System Operation

To develop an efficient PV panel cooling system, a computational code was developed in the MATLAB environment to generate a finite volume mesh enabling optimization of the PV module cooling system (Figs. 7.8 and 7.9). The mathematical model of the PV cooling system was also implemented in the MATLAB package. The program enables both optimization of the cooling system geometry and transient-state calculations making it possible to determine the efficiency of the system operation.

The analyses are based on the assumption of the three-dimensional heat transfer in the PV panel and the one-dimensional heat transfer in the cooling medium, as described in this chapter.

The analysed computational cases are listed in Table 7.1.

The analysis is based on the model parameters presented in Table 7.2.

The thermophysical properties of the 50/50 water–glycol mixture (dependent on the fluid temperature) are described by the following quantities expressed as:

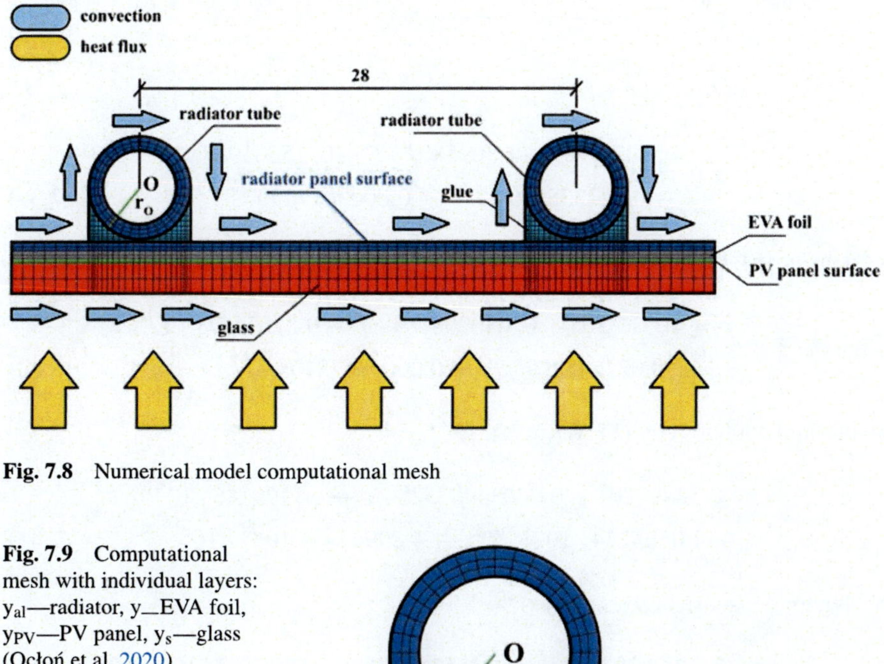


**Fig. 7.8** Numerical model computational mesh

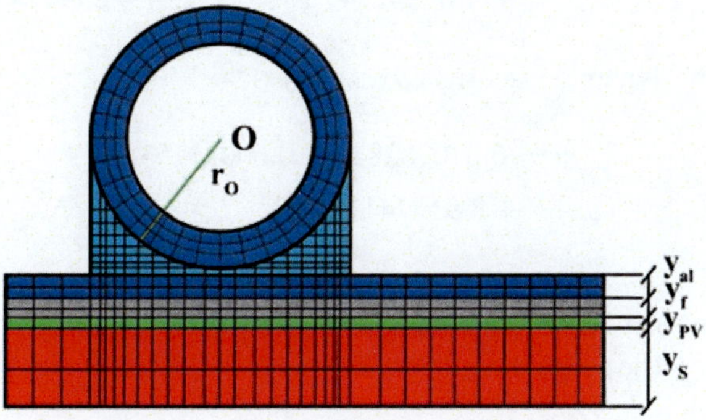


**Fig. 7.9** Computational mesh with individual layers: $y_{al}$—radiator, y_EVA foil, $y_{PV}$—PV panel, $y_s$—glass (Ocłoń et al. 2020)

**Table. 7.1** Computational cases under analysis

| | $G$(W/m$^2$) | $h_e$ (W/(m$^2$ K)) | $T_{in}$ (°C) | $T_{ext}$ (°C) |
|---|---|---|---|---|
| 1 | 1,100 | 10 | 20 | 35 |
| 2 | 700 | 10 | 18 | 22 |
| 3 | 500 | 10 | 15 | 18 |
| 4 | 300 | 8 | 12 | 14 |
| 5 | 200 | 15 | 12 | 10 |

**Table. 7.2** Numerical model parameters adopted for the calculations

| | $k$ (W/(m K)) | $c$ (J/(kg K)) | $\rho$ (kg/m$^3$) | Layer thickness (mm) |
|---|---|---|---|---|
| PV panel | 400 | 670 | 2,330 | 0.5 |
| EVA film | 0.34 | 1,400 | 1,000 | 1 |
| Glass layer | 0.8 | 670 | 2,800 | 3 |
| Radiator | 200 | 900 | 2,700 | 1 |
| Radiator tube | 200 | 900 | 2,700 | $r_{in} = 3$ mm, $r_{out} = 4$ mm |

- specific heat ($c$), J/(kg·K):

$$c = 3447.0929 + 1.3095802 \cdot T + 0,15361977 \cdot T^2 - 0.022338179 \cdot T^{2.5} + 0.00090170907 \cdot T^3 \tag{7.39}$$

- density (**ρ**), kg/m$^3$:

$$\rho = 1051.5565 + 0,030908084 \cdot T - 0.21676671 \cdot T^{1.5} + 0.024545809 \cdot T^2 + 0.00098678108 \cdot T^{2.5} \tag{7.40}$$

- thermal conductivity ($k$), W/(m K)

$$k = 0.38492201 + 0.00034482528 \cdot T - 6.1199146 \cdot 10^{-6} \cdot T^2 - 1.0565274 \cdot 10^{-6} \cdot T^{2.5} - 4.8994243 \cdot 10^{-8} \cdot T^3 \tag{7.41}$$

- dynamic viscosity ($\mu$), kg/(m·s)

$$\mu = 0.010982847 - 0.0002375343 \cdot T + 1.4800407 \cdot 10^{-5} \cdot T^{1.5} + 3.9127417 \cdot 10^{-6} \cdot T^2 - 1.0996693 \cdot 10^{-8} \cdot T^3 \tag{7.42}$$

Considering the system symmetry, only a half of the PV panel with the dimensions of 0.49 m × 1.64 m is modelled.

The four cases under analysis include different values of:

1. the heat flux (solar radiation intensity) incident on the panel surface,
2. the cooling medium inlet temperature,
3. the ambient temperature.

The analysis covers:

1. the heat flux absorbed by the cooling medium,
2. the panel maximum temperature,
3. the temperature difference between the glass surface and the radiator,
4. the inlet/outlet temperature difference (for the water–glycol mixture).

The analysis results are presented in Figs. 7.10, 7.11, 7.12, 7.13 and 7.14.

Analysing Figs. 7.10, 7.11, 7.12, 7.13 and 7.14, it can be observed that the cooling system operates correctly both at a low ambient temperature and solar radiation intensity (Case 4) and at a high outdoor temperature and solar radiation intensity. On a cloudy day with medium solar radiation intensity of 300 W/m$^2$ it is possible to obtain a heat flux of up to 200 W (Case 4); on a sunny day and the solar radiation

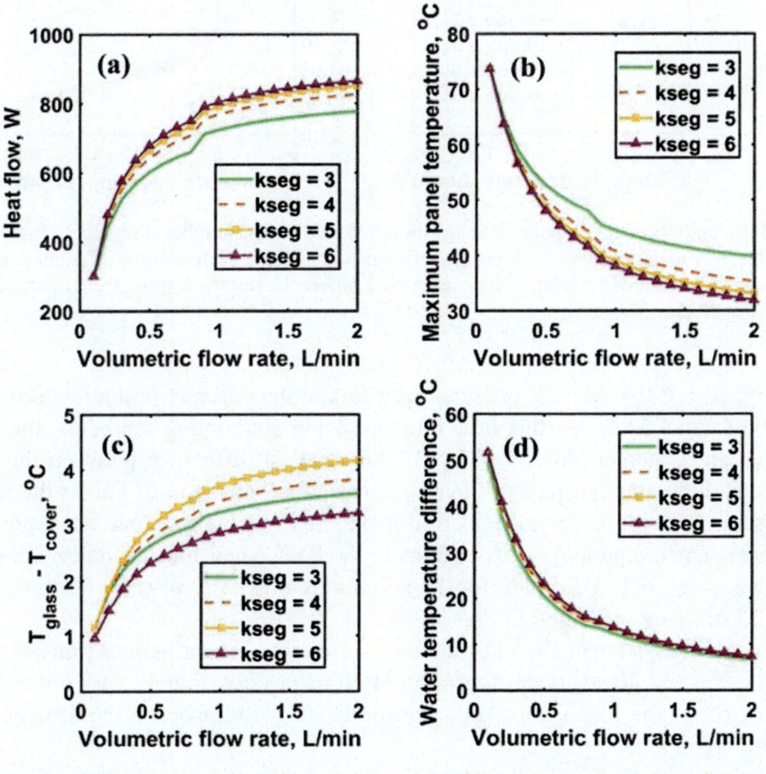


**Fig. 7.10** Analysis of the PV panels cooling process for a different number of segments: kseg = 3, 4, 5, 6: **a** heat flux absorbed by water, **b** panel maximum temperature, **c** temperature difference between the glass and the radiator, **d** inlet/outlet temperature difference (for the water–glycol mixture). Tube pitch: $L_T = 28$ mm (Case 1)

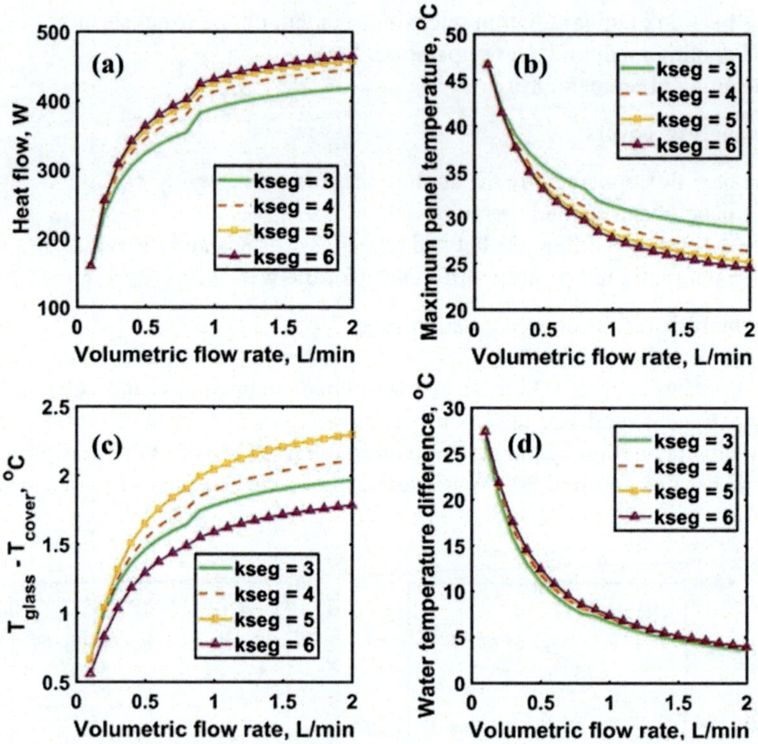


**Fig. 7.11** Analysis of the PV panels cooling process for a different number of segments: kseg = 3, 4, 5, 6: **a** heat flux absorbed by water, **b** panel maximum temperature, **c** temperature difference between the glass and the radiator, **d** inlet/outlet temperature difference (for the water–glycol mixture. Tube pitch: $L_T = 28$ mm (Case 2)

intensity of 1,100 W/m$^2$ the obtained heat flux at the volumetric flow rate of $Q_v =$ 2 L/min totals 850 W. At this flow rate and using six cooling segments, the panel surface can be cooled down to 32 °C. The smallest difference between the glass surface and the outer surface of the radiator occurs for $kseg = 6$. This is the reason why the ELFRAN company decided to make the PV panel cooling system using 6 segments. The temperature distributions in the PVT panel for different numbers of segments ($kseg = 3$, 4, 5 and 6) for the volumetric flow rates of $Q_v = 0.5$ L/min and $Q_v = 2$ L/min are presented in Fig. 7.15.

The cooling system mathematical model was verified experimentally on 6 October 2018 on the test stand made under the HySOL project. Figure 7.16 presents the distribution of the panel inlet ($T_{in}$) and outlet ($T_{out}$) temperature and of outdoor temperature ($T_{ext}$).

The curve in Fig. 7.17 illustrates the changes in the solar radiation intensity implemented for the calculations.

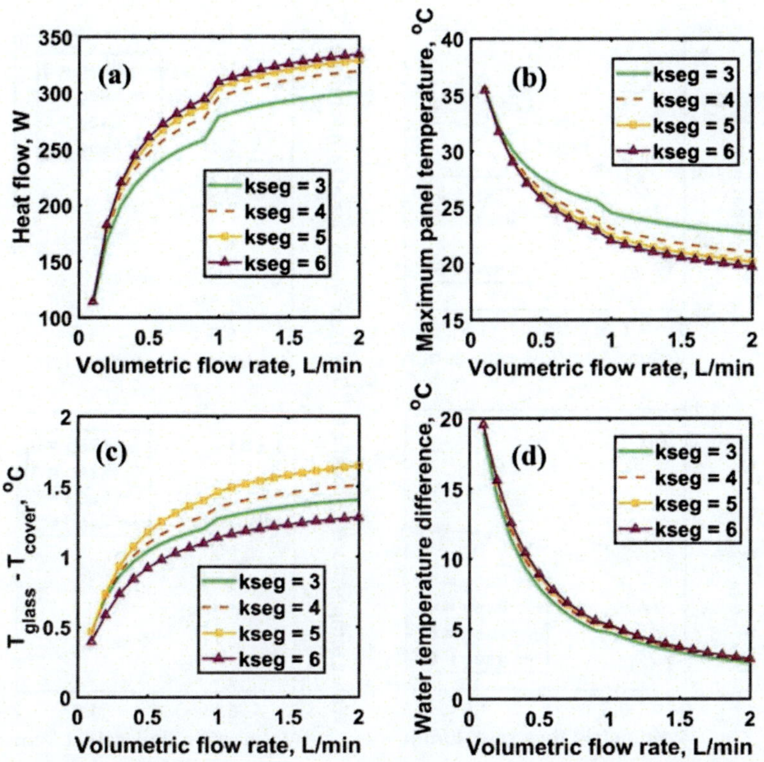


**Fig. 7.12** Analysis of the PV panels cooling process for a different number of segments: kseg = 3, 4, 5, 6: **a** heat flux absorbed by water, **b** panel maximum temperature, **c** temperature difference between the glass and the radiator, **d** inlet/outlet temperature difference (for the water–glycol mixture.) Tube pitch: $L_T = 28$ mm (Case 3)

Figure 7.18 presents the history of changes in the wind speed. Wind speed is taken into account in numerical calculations of the heat transfer coefficient $h_e$ (Equation 7.9).

The parameter that has the key impact on the cooling efficiency is the volumetric flow rate. The changes in this quantity recorded on 6 October 2018 are presented in Fig. 7.19. The volumetric flow rate value was selected so that a possibly constant difference could be maintained between the cooling system inlet and outlet temperatures in the period between 9 a.m. and 5 p.m. (during the solar radiation strongest intensity).

Figure 7.20 presents the results obtained from the calculations using the developed mathematical model of the PV panel cooling system. The history of the panel outlet temperature values obtained numerically are compared with the experimental testing results.

The calculations were performed for three different numerical meshes. The number of the mesh elements in the domain of the PV panel and the cooling system

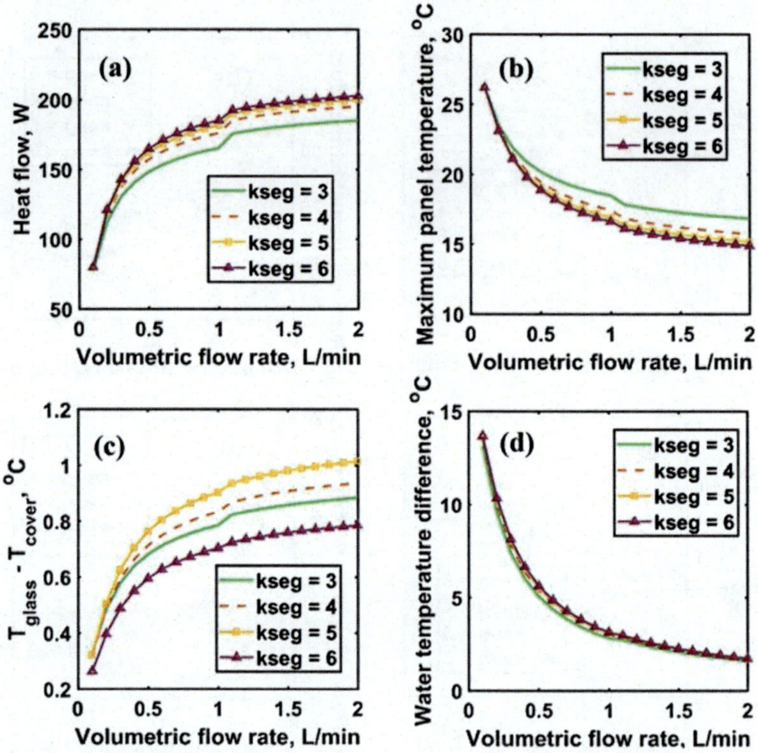


**Fig. 7.13** Analysis of the PV panels cooling process for a different number of segments: kseg = 3, 4, 5, 6: **a** heat flux absorbed by water, **b** panel maximum temperature, **c** temperature difference between the glass and the radiator, **d** inlet/outlet temperature difference (for the water–glycol mixture). Tube pitch: $L_T = 28$ mm (Case 4)

varied in the range of 15,810–34,782, whereas the number of nodes—in the range from 21,492 to 42,984. Time step $\Delta\tau$ adopted for the calculations varied from 0.546 s to 1.2 s (Table 7.3).

The obtained histories of the temperature values are presented in Fig. 7.21. It can be seen clearly that the differences between the obtained histories of the outlet temperature values are slight and do not exceed 1 °C. Mesh 3 was selected for the calculations.

Figure 7.22 presents the temperature distribution on the surface of the cooled PV panel at 3 p.m., when the solar radiation intensity reached almost 700 W/m$^2$ (cf. Fig. 7.16). The cooled panel temperature is slightly above 30 °C. The temperature difference between the cooling medium inlet and outlet is 11 °C. The cooling system operates very efficiently—the temperature difference between the glass surface and the surface of the radiator tubes does not exceed 2 °C.

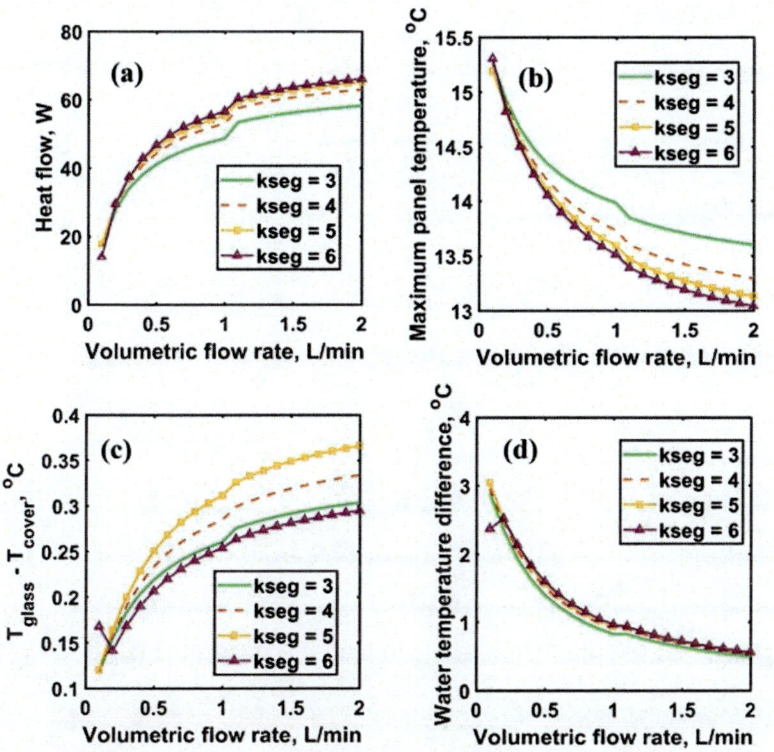


**Fig. 7.14** Analysis of the PV panels cooling process for a different number of segments: kseg = 3, 4, 5, 6: **a** heat flux absorbed by water, **b** panel maximum temperature, **c** temperature difference between the glass and the radiator, **d** inlet/outlet temperature difference (for the water–glycol mixture). Tube pitch: $L_T = 28$ mm (Case 5)

The conclusions that can be drawn from the analysis of Fig. 7.20, which illustrates the comparison between the results of experimental testing (measurement of the cooling medium outlet temperature) and numerical simulations, are as follows:

1. the model reflects well the changes in the outlet temperature during the period of high solar radiation (10 a.m. to 3 p.m.);
2. in the period from 8 a.m. to 10 a.m., the model responds too slowly to accurately capture changes in the cooling medium outlet temperature;
3. in the period from 4 p.m. to 6 p.m., the model predicts slightly higher temperature values than those obtained from measurements.

Despite some imperfections of the model, the agreement between the results of calculations and numerical simulations is satisfactory.

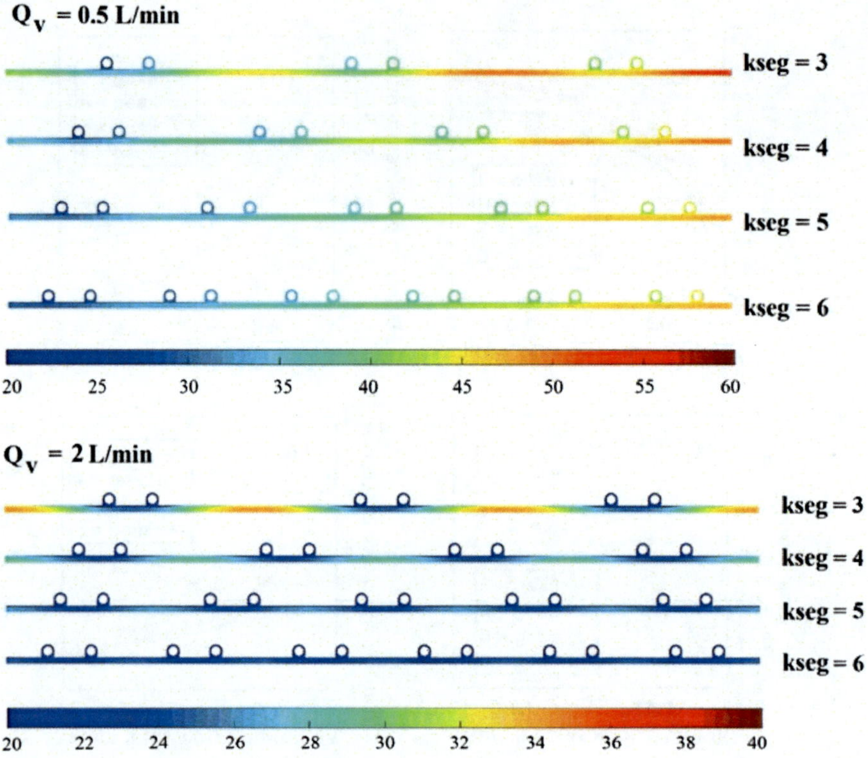


**Fig. 7.15** Temperature distributions in the PVT panel for kseg = 3, 4, 5 and 6 and for the volumetric flow rates of $Q_V = 0.5$ L/min and $Q_V = 2$ L/min

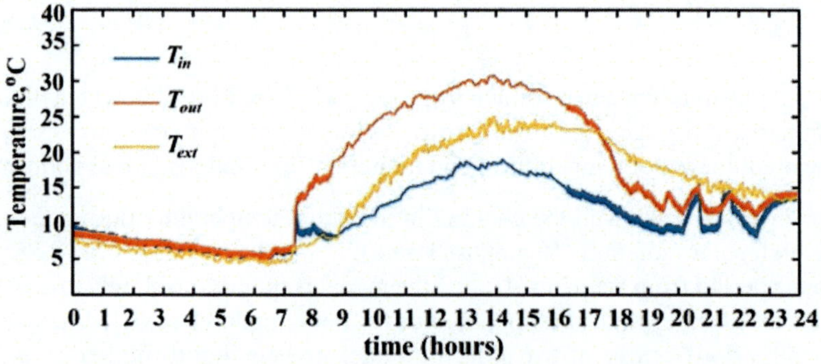


**Fig. 7.16** Histories of the inlet ($T_{in}$) and outlet ($T_{out}$) temperature of the PV panel with the cooling system and variations in the ambient temperature ($T_{ext}$) on 6 October 2018

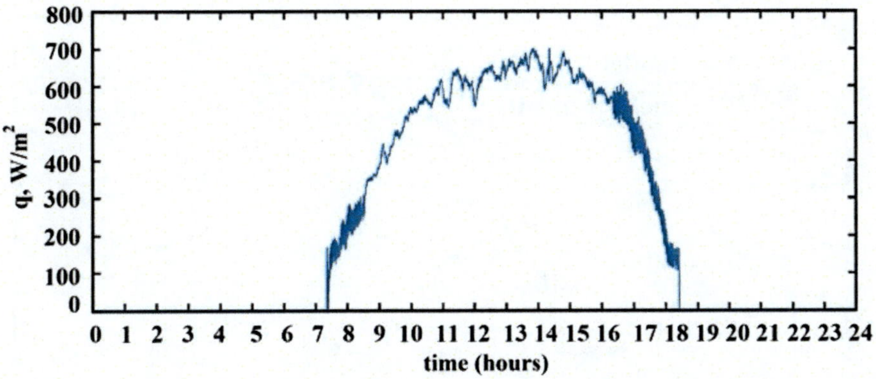


**Fig. 7.17** Changes in the solar radiation intensity for the PV panel equipped with the sun-tracker on 6 October 2018

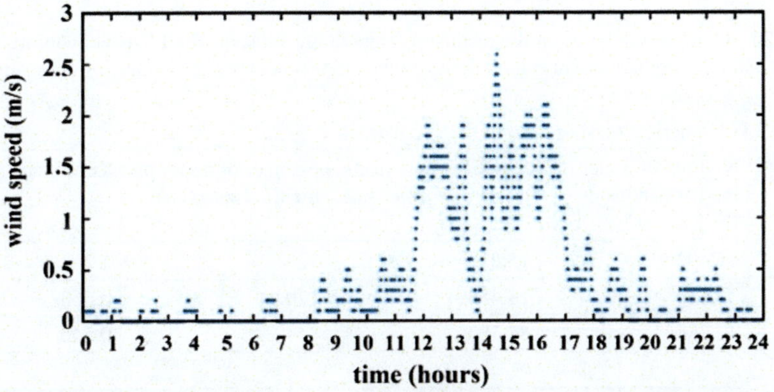


**Fig. 7.18** Values of wind speed $u_e$ measured by the weather station on 6 October 2018

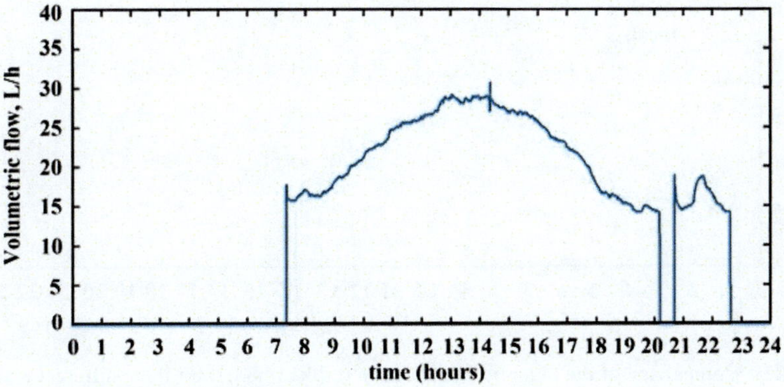


**Fig. 7.19** Changes in the values of the volumetric flow rate $Q_v$ during the testing of the PV panel cooling system on 6 October 2018

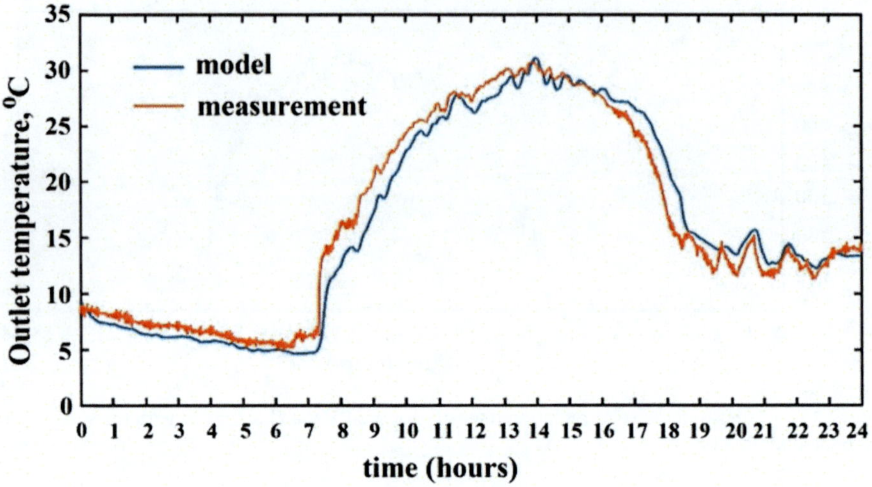


**Fig. 7.20** Comparison between the results of temperature values at the PV panel cooling system outlet obtained from numerical calculations (model) and from experimental testing (measurement)

**Table. 7.3** Numerical meshes used in the simulations

| Mesh | Number of nodes on the PV panel side ($N_s$) | Number of elements on the PV panel side | Number of nodes on the fluid side ($N_f$) | Time step ($\Delta\tau$), s |
|---|---|---|---|---|
| Mesh 1 | 21,492 | 15,810 | 72 | 1.2 |
| Mesh 2 | 35,820 | 28,458 | 120 | 0.668 |
| Mesh 3 | 42,984 | 34,782 | 144 | 0.546 |

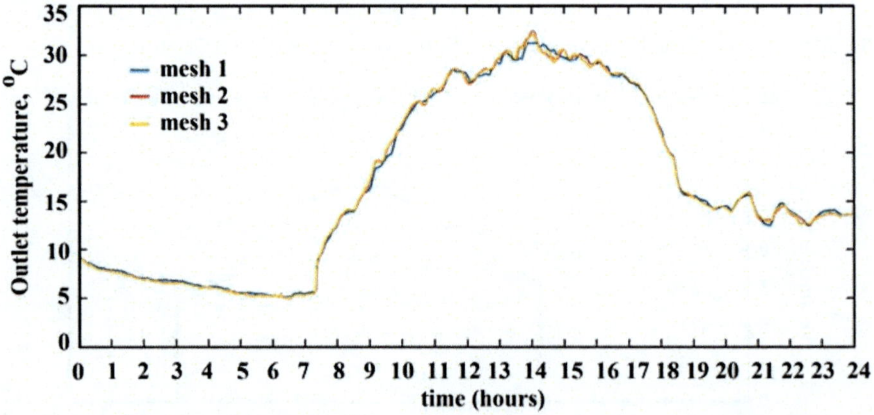


**Fig. 7.21** Comparison of the numerical calculation results (model) for three different numerical meshes (mesh 1, mesh 2 and mesh 3; given in Table 7.3)

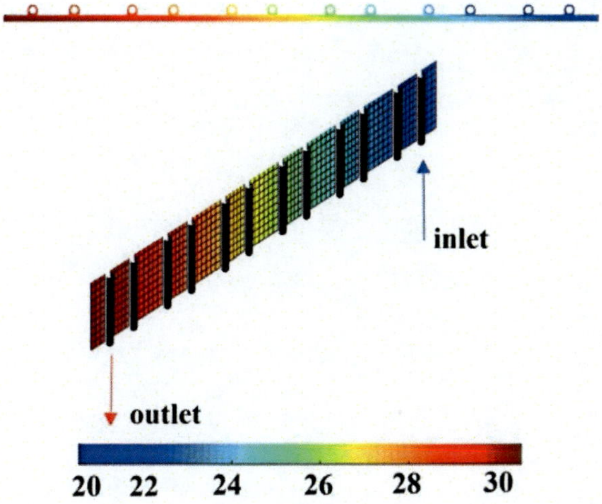


**Fig. 7.22** Temperature distribution on the surface of the cooled PV panel (3 p.m.)

# References

Kozak-Jagieła E (2020), PhD Thessis: „Modeling and experimental investigation of heat transfer for a new active cooling system for photovoltaic panels" (in polish)

Ocłoń P, Cisek P, Kozak E, Taler J, Taler D, Skrzyniowska D, Fedorczak-Cisak M (2020) Modeling and experimental validation and thermal performance assessment of a sun-tracked and cooled PVT system under low solar irradiation. ENERGY CONVERS. MANAG 222:113289

Taler D, Ocłoń P (2014) Thermal contact resistance in plate fin-and-tube heat exchangers, determined by experimental data and CFD simulations. Int J Therm Sci 84:309–322

# Chapter 8
# Economic Analysis

The cost-effectiveness of the RESHeat system intended for residential building heating was evaluated based on the dynamic method using discounted profit and value methods, such as:

- NPV—net present value,
- NPVR—net present value rate,
- IRR—internal rate of return,
- DPB—discounted payback period,
- SPB—simple payback period.

The NPV method is one of the dynamic methods of calculating the profitability account of a given investment. It makes it possible to take account of all future inflows and outflows based on the present value of inputs.

The net present value (NPV) is the value obtained by discounting in every year the difference between inflows and outflows over the entire period, assuming a constant discount rate. The NPV level depends on the volume and kind of net cash flows in a given period and on the average discount rate. If NPV > 0, the project is accepted; if not (NPV < 0) – it is rejected. Finally, if NPV = 0, it is impossible to decide on the project. The NPV is calculated from formula (8.1):

$$NPV = \sum_{i=1}^{n} \frac{CF_i}{(1+r)^n} - J_o \rightarrow [€] \tag{8.1}$$

where.
$CF$—cash flows.
$n$—time period under analysis.
$r$—discount rate.
$J_o$—investment costs.
$CF_i$ = profit – loss.



P. Ocłoń, *Renewable Energy Utilization Using Underground Energy Systems*, Lecture Notes in Energy 84, https://doi.org/10.1007/978-3-030-75228-6_8

Cash flows include all the flows that will be generated by the analysed investment in the future.

The NPVR value is the relationship between the NPV and the present (updated) value of the required investment outlay, which is expressed by Eq. (8.2):

$$\text{NPVR} = \frac{\text{NPV}}{J_o} \tag{8.2}$$

An important quantity often taken into account in economic analysis is the internal rate of return (IRR).

The IRR is the rate of return when the NPV is zero (NPV = 0).

$$0 = \text{NPV} = \sum\nolimits_{i=1}^{n} \frac{CF_i}{(1 + IRR)^n} - J_o \tag{8.3}$$

An investment is cost-effective if the IRR is higher than the limit rate of return, which is the lowest rate of return acceptable to the investor. The limit rate of return is usually the interest rate on long-term loans or paid by any borrower. In other words, if the IRR is higher than the cost of raising capital, the safety margin and the project profitability are bigger, which means a bigger difference between the IRR and the limit rate of return.

The value of the simple payback period (SPB) defines the time after which the investment will provide at least a return on investment expenditures under the assumption that $CF_t = const$.

$$\text{SPB} = \frac{J_o}{CF_t} \tag{8.4}$$

The value of the discounted payback period (DPB) is the number of periods after which the sum of discounted net cash flows is zero.

$$\text{DPB} = \frac{\ln(1 - \text{r} \cdot \text{SPB})}{-\ln(1 + \text{r})} \tag{8.5}$$

This work draws a comparison between heating systems of two buildings constructed in 1970. The buildings are poorly insulated (10 cm of expanded polystyrene) with single-pane windows and a flat roof. There are 50 apartments in the first building and 20 in the second (Table 8.1).

The proposed heating option is the RESHeat system using renewable energy sources.

The capital cost of the RES-based option is shown in Table 8.2. The table presents the prices of the system individual components.

The RESHeat system was compared to the following heating options for the example building:

– heating using a condensing boiler fired with natural gas

Table 8.1 Current costs of heating for buildings with 50 and 20 apartments

| Month | 50 apartments | | 20 apartments | |
|---|---|---|---|---|
| | Thermal energy (GJ) | Cost (€) | Thermal energy (GJ) | Cost (€) |
| October | 128.00 | 784.58 | 51.2 | 313.83 |
| November | 83.80 | 510.24 | 33.52 | 204.10 |
| December | 465.60 | 2836.17 | 186.24 | 1134.47 |
| January | 355.00 | 2186.89 | 142.00 | 874.75 |
| February | 256.00 | 1577.02 | 102.40 | 630.81 |
| March | 181.60 | 1118.70 | 72.64 | 447.48 |
| April | 129.60 | 798.37 | 51.84 | 319.35 |
| May | 73.00 | 456.00 | 29.20 | 182.40 |
| Fixed connection fee | | 7293.67 | | 4107.19 |
| Total | 1672.60 | 17,561.65 | 669.04 | 7754.03 |

Table 8.2 Cost of the RESHeat system for 50 and 20 apartments

| Cost of the RESHeat system | 50 apartments | 20 apartments |
|---|---|---|
| | Cost (€) | Cost (€) |
| Heat pump (250 kW) | € 40,000.00 | € 20,000.00 |
| Boreholes and vertical heat exchangers | € 22,000.00 | € 12,000.00 |
| PVT modules (65 kW) | € 78,000.00 | € 31,200.00 |
| Piping and automatic control | € 12,000.00 | € 8,000.00 |
| Accumulation tank | € 10,000.00 | € 6,000.00 |
| Subtotal | **€ 162,000.00** | **€ 77,200.00** |
| Donation (20%) | € 15,600.00 | € 6,240.00 |
| Total cost ($J_o$) | **€ 146,400.00** | **€ 70,960.00** |

- heating from the district heating network
- heating using a boiler fired with biomass
- heating using a ground source heat pump

Tables 8.3 and 8.4 show investment and operating costs for all the analysed heating options of the building with 50 and 20 apartments, respectively.

As indicated by Table 8.4, the investment costs are the lowest for gas heating and the highest for the RESHeat system. However, looking at the annual costs of energy, gas heating is the most expensive option compared to heating with biomass or heating from the district heating network (Table 8.5).

It is assumed that the RESHeat system operating costs are zero because the electricity consumed by the heat pump is supplied entirely by the photovoltaic installation. The system requires no additional annual investment outlays, which is a significant advantage compared to other heating methods. The highest heating costs in the

**Table 8.3** Investment and operating costs of the building with 50 apartments

| | Investment costs ($J_o$) | Operating costs |
|---|---|---|
| RESHeat | € 146,400.00 | 0 |
| Biomass | € 21,000.00 | € 20,124.87 |
| District heating network | € 23,800.00 | € 17,561.65 |
| Natural gas | € 8000.00 | € 20,907.50 |
| Ground source heat pump | € 74,000.00 | € 19,900.00 |

**Table 8.4** Investment and operating costs of the building with 20 apartments

| | Investment costs ($J_o$) | Operating costs |
|---|---|---|
| RESHeat | € 70,960.00 | 0 |
| Heating with biomass | € 10,500.00 | € 8049.95 |
| District heating network | € 20,400.00 | € 7754.00 |
| Natural gas | € 4700.00 | € 8363.00 |
| Ground source heat pump | € 40,000.00 | € 7960.00 |

**Table 8.5** Values of individual parameters of the financial analysis for 50 apartments

| | Biomass | Gas | District heating network | Heat pump |
|---|---|---|---|---|
| NPV | € 59,267 | € 40,900 | € 23,710 | € 85,406 |
| NPVR | 0.47 | 0.30 | 0.19 | 1.18 |
| SPB | 6.75 | 7.68 | 8.35 | 4.54 |
| DPB | 7.32 | 8.43 | 9.23 | 4.81 |
| IRR | 0.10 | 0.07 | 0.06 | 0.21 |

presented case are for gas heating. The annual costs for each option are in the range of 20,000 € for 50 apartments and about 8000 € for 20 apartments.

The presented analysis assumes a constant discount rate of $r = 2$.

Figure 8.1 shows the net present value (NPV) over the period of consecutive years for the building with 50 apartments. The RESHeat system will provide the fastest return compared to heating with ground source heat pump. The return will be the latest for the option with heating from the district heating network.

Figure 8.2 presents the net present value (NPV) over the years for the building with 20 apartments.

Table 8.6 presents the values of individual parameters of the financial analysis for the heating options compared to the RESHeat system (for 20 apartments). It can clearly be seen that choosing the RESHeat option is favourable. Despite the fact

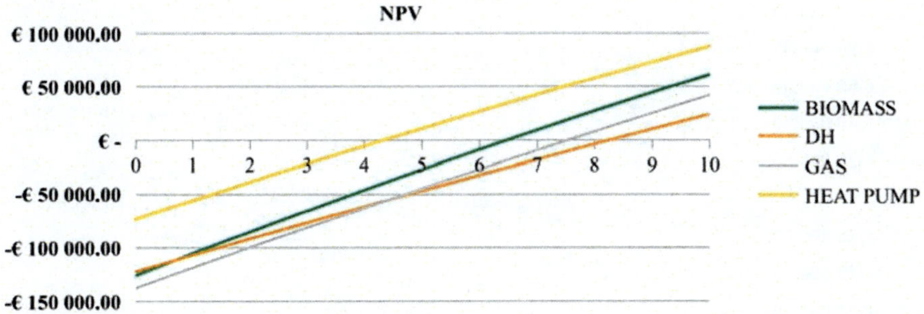


**Fig. 8.1** NPV over the years for 50 apartments

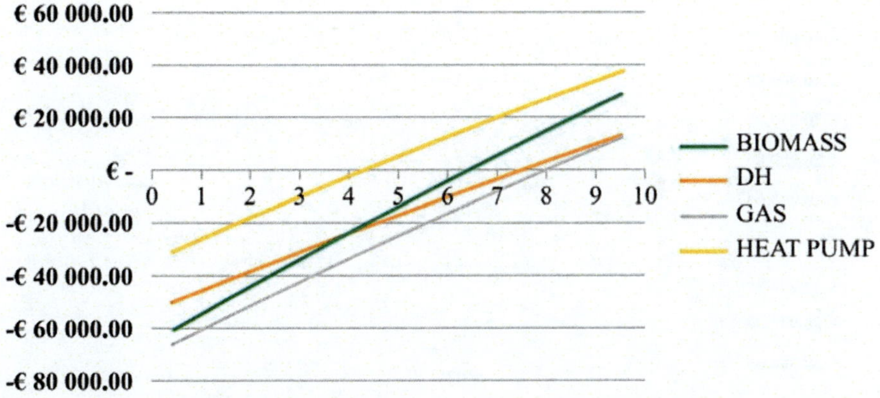


**Fig. 8.2** NPV over the years for 20 apartments

**Table 8.6** Values of individual parameters of the financial analysis for the building with 20 apartments

| | Biomass | Gas | District heating network | Heat pump |
|---|---|---|---|---|
| NPV | € 27,900 | € 13,255 | € 11,950 | € 36,675 |
| NPVR | 0.46 | 0.26 | 0.18 | 1.18 |
| SPB | 6.80 | 7.89 | 8.45 | 4.53 |
| DPB | 7.38 | 8.67 | 9.34 | 4.79 |
| IRR | 0.10 | 0.07 | 0.05 | 0.21 |

that the investment cost is quite high, the cost of the RESHeat system installation is returned.

The SPB period for biomass is almost 7 years, for gas—almost 8 years, for the district heating network option—about 8.5 years, whereas for the heat pump the SPB period totals about 4.5 years.

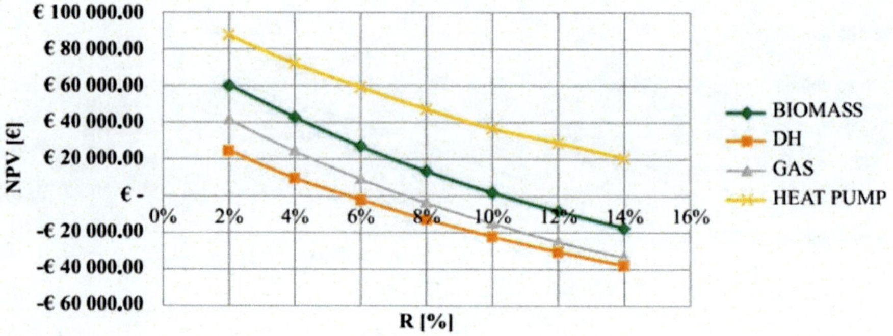


**Fig. 8.3** NPV profile depending on the discount rate for 20 apartments

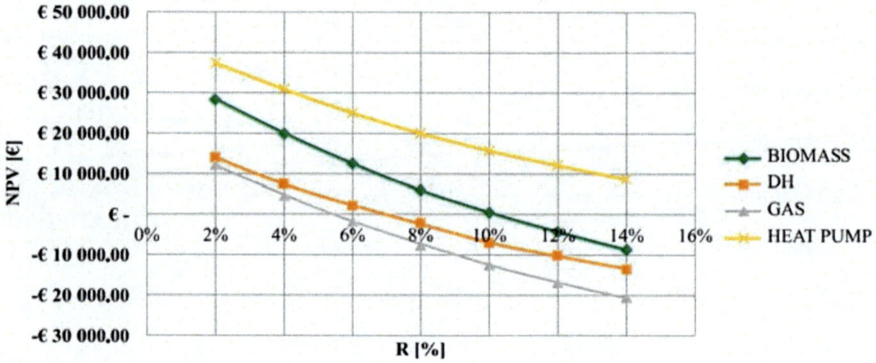


**Fig. 8.4** NPV profile depending on the discount rate for 50 apartments

Figures 8.3 and 8.4 show how the NPV profile changes depending on the value of the discount rate. The higher the discount rate, the bigger the decrease in the net present value.

Figure 8.3 presents the NPV profile as a function of the discount rate for 20 apartments. For 20 apartments, the RESHeat system with a heat pump is cost-effective even at r = 14%. In the case of gas heating, the investment loses profitability at $r$ higher than 5%, for biomass—at $r$ higher than 7% and for district heating—at $r$ exceeding 10%.

For 50 apartments and compared to heating based on natural gas, the RESHeat option (Fig. 8.4) ensures a return at $r < 5\%$, compared to district heating—at $r < 7\%$ and for heating using a biomass-fired boiler—at $r < 10\%$.

# Chapter 9
# Advantages of the Resheat System Over Traditional Solutions

Table 9.1 presents the advantages of the RESHeat system over traditional solutions used for heating purposes.

**Environmental benefits**

As already mentioned, the RESHeat system is a zero-emission system of the building heating and cooling. The environmental benefits are determined using the emission coefficients presented in Table 9.2.

South-eastern Poland (Krakow, Limanowa) is one of the most polluted regions in the EU due to the use of conventional fuels (coal, biomass, natural gas). For this reason, RES-based solutions for heating and cooling are essential for the health of the region inhabitants (Table 9.3).

In the case of the two buildings under analysis in Chap. 8, due to the application of the RESHeat system, the reduction in $CO_2$ emissions totals 93.79 tons per year for the building with 50 apartments and 37.51 tons for the building with 20 apartments. Comparing the RESHeat system to a natural gas boiler, additional reductions of 90 and 36 kg of CO and 66.88 and 26.75 kg of $SO_x$ are achieved, respectively. Compared to biomass heating, the application of the RESHeat system causes a reduction of 71.89 kg and 28.75 kg of $NO_x$, 90 and 36 kg of CO and 8.36 kg and 3.34 kg of PM for the building with 50 and 20 apartments, respectively.



P. Ocłoń, *Renewable Energy Utilization Using Underground Energy Systems*, Lecture Notes in Energy 84, https://doi.org/10.1007/978-3-030-75228-6_9

**Table 9.1** Advantages of the RESHeat solution compared to heating based on different heat sources

| Traditional heating systems | Advantages of the RESHeat solution compared to conventional heating systems |
|---|---|
| Boiler fired with biomass | (a) The RESHeat solution is a zero-emission system intended for heating buildings, whereas biomass combustion results in $NO_x$ emissions, which are harmful to the environment<br>(b) Biomass needs to be stored, which requires a storage space. This generates additional costs<br>(c) The cost of biomass transportation is higher compared to the cost of the RESHeat system maintenance<br>(d) Biomass combustion is prohibited in many urban areas<br>(e) The RESHeat system is based on the use of heat pumps. Therefore, the service life of the system is much longer (25 years) compared to boilers firing biomass (10 years) |
| Boilers fired with natural gas | (a) Longer service life (25 years for the heat pump, 10 years for natural gas boilers)<br>(b) No greenhouse gas emissions |
| Ground source heat pumps | (a) Advanced ground regeneration, which makes it possible to keep the COP at a high level in consecutive seasons of the heat pump operation<br>(b) A higher COP due to ground regeneration through utilization of waste heat from the PVT panels<br>(c) No operating costs |
| District heating networks | (a) Due to emission taxes, the cost of domestic hot water preparation will increase in Poland by about 30% in the next 5 years, so the payback period of RES investments compared to district heating may decrease from 7.8 to 6 years<br>(b) Not all buildings in the city can be supplied from the district heating network, while the use of the RESHeat system is not limited by the network availability<br>(c) District heating and cogeneration (CHP generation) are mainly based on fossil fuels, which involves GHG emissions. The RESHeat system is based exclusively on renewable energy sources; it is a system of energy production causing no GHG emissions |

**Table 9.2** Emission coefficients adopted in the environmental analysis

| | Emission coefficients | | | | |
|---|---|---|---|---|---|
| | $CO_2$ kg/GJ | CO kg/GJ | $SO_x$ kg/GJ | $NO_x$ kg/GJ | PM kg/GJ |
| Natural gas | 56.1 | 0.054 | 0.0 | 0.040 | 0.0 |
| Biomass | 0.0 | 0.054 | 0.007 | 0.043 | 0.005 |

**Table 9.3** Reduction in $CO_2$, CO, $SO_X$, $NO_X$, and PM emissions per year

| | $CO_2$, t | CO, kg | $SO_X$, kg | $NO_X$, kg | PM, kg |
|---|---|---|---|---|---|
| | *Residential building—50 apartments* | | | | |
| Gas | 93.799 | 90.192 | 0.0 | 66.880 | 0.0 |
| Biomass | 0.0 | 90.192 | 11.704 | 71.896 | 8.360 |
| | *Residential building—20 apartments* | | | | |
| Gas | 37.52 | 36.077 | 0.0 | 26.752 | 0.0 |
| Biomass | 0.0 | 36.077 | 4.681 | 28.758 | 3.344 |

# Chapter 10
# Optimization of Underground Power Cable Systems

The continued development of the world economy and population growth, especially in developing countries, entail an increase in the demand for electricity supply. Electricity around the world is transmitted mainly by high voltage (HV) alternating current (AC) overhead line technology. For example, 96% of the overhead transmission network in Europe is built overhead and only 4% runs underground. Underground cables are mainly used for short distances, in places where overhead lines are not advisable or cannot be used, and for special technical applications. The use of underground cabling is becoming increasingly attractive mainly for environmental and aesthetic reasons. Also, underground transmission lines (UTL) are weatherproof and are installed when the operation of overhead lines may cause adverse environmental impacts, concerns about potential health problems, and impacts on property prices or the condition of national parks or areas of natural beauty. UTLs are also more reliable than overhead transmission lines in terms of the likelihood of cable line failure. Therefore, UTL is recommended for design and installation in:

- densely populated urban areas (easy grid expansion, less risk of electrocution),
- power generation by power plants (i.e., conventional or renewable energy sources) and large energy consumers (i.e., mines, steel mills, manufacturing plants, and even interconnections between countries),
- interconnections at power plants.

However, the use of underground cables in HV applications is still limited due to their high cost and the cost of maintenance and repair during failures. It was estimated that underground lines can take 48–480 h to repair during a fault, compared to 8–48 h for overhead lines. Underground cables themselves also have a higher unit price due to their structural complexity and higher production costs (Metwally et al. 2013).

Undergrounding is not a new trend, as underground cables have long been used in low and medium voltage lines in urban areas. Several countries use underground low-voltage (LV) distribution lines in almost all of their networks, with a goal of 100%, such as the Netherlands, Singapore, and Denmark. Denmark, for example, plans to underground 75% of its electricity grid soon. Note that almost 10% of the planned



P. Ocłoń, *Renewable Energy Utilization Using Underground Energy Systems*, Lecture Notes in Energy 84, https://doi.org/10.1007/978-3-030-75228-6_10

underground power lines are 400 kV high voltage transmission lines (ENERGETIK DK 2009).

The designer of an underground HV transmission line faces difficulties in dealing with UTL factors that affect the reliability and investment cost of the underground cable system. An important factor limiting the current carrying capacity of a power line is the cross-sectional area of the conductor of the feeder cable. The current carrying capacity of a conductor is defined as the maximum electrical current that the conductor can safely carry without exceeding the insulation temperature limitations. Thus, the larger the cross-sectional area of the cable conductor, the higher the current that can be safely transmitted under the given conditions. The current carrying capacity of a cable conductor depends mainly on the temperature of the cable core. Therefore, the greater the electric current that is passed through the cable conductor, the greater the amount of heat generated as Joule heat in the cable core. The heat generated can cause the cable core temperature to increase under adverse heat dissipation conditions. Thus, exceeding cable conductor temperature above 90 °C for XLPE insulated power cables leads to overheating of the cable and malfunction of the power line. Excessive cable core temperature for a long time can result in melting of polyethylene insulation, which in turn will cause transmission line failure. The repair time for underground HV transmission lines is about ten times longer than for overhead lines. Since an outage can last as long as 480 h, and each hour entails a huge financial loss due to interruption of power transmission, the failure rate of HV UTLs is required to be reduced to a minimum.

Another major barrier to the use of underground HV cables is the dissipation of heat from the cable duct to the environment. Cable systems buried directly in native soil face almost unpredictable variability in environmental conditions (changes in soil type, thermal properties of the soil along the line). According to the "chain" principle—chain is only as strong as its weakest link—the most vulnerable part of the cable line must be identified when designing the UPCL. Therefore, it is important to determine the most unfavourable conditions for burial of the considered cable line, e.g. sections of the line where the maximum allowable temperature of the cable core can be exceeded.

To evaluate the performance of an underground cable line, the thermal properties of the soil in the vicinity of the cable line must be known. Due to significant changes in geological conditions (changes in soil layering, soil composition, and water content) along the cable line route, and thus changes in the thermal properties of the native soil, it is not possible to determine the constant heat transfer. conditions between the cables and the surrounding soil. Therefore, an effective backfill material must be used to ensure the reliability of the line by providing favourable heat dissipation conditions in the extensive vicinity of the cable. A suitable backfill composition provides better heat dissipation conditions compared to the native soil. Consequently, the current carrying capacity of the UTL can be increased by about 10 to 15%. This can also result in a reduction in the cross-sectional area of the cable conductor, thereby reducing investment costs. Another advantage of a suitable backfill material is the stability of its thermal properties over time, ensuring reliable operation of the transmission line over the lifetime of the cable system. Therefore, by providing favourable thermal conditions through the use of backfill materials, adverse weather conditions such as soil drying will have limited impact on the performance of the transmission line. Therefore, the increased current carrying capacity will be maintained throughout the life of the system.

Increasing power supply requirements and improvement of transmission line reliability necessitate optimization of power systems. The optimization procedure should result in an appropriate design of the UTL with favorable burial conditions. In this way, a significant increase in the current carrying capacity of the cable under given conditions can be achieved. Thus, it will improve the flexibility of the transmission line and ensure its safety operation. Furthermore, the improved heat dissipation conditions can lead directly to a reduction in the cross-sectional area of the cable conductor, thus enabling a significant reduction in the installation cost of the UPCL without changing the current rating carried.

Several attempts have already been made to optimize underground power cable systems (UPCS). Del Valle et al. (2008) presented a detailed review of the basic concepts of the Particle Swarm Optimization (PSO) algorithm and its variants and delivered a comprehensive overview of power system applications. This analysis demonstrated the potential of the PSO algorithm in UPCS optimization.

Kovač et al. (2006) proposed a numerical-stochastic technique developed for UPCS optimization. The method presented was based on nonlinear electrothermal modelling of interconnected underground cables and a stochastic optimization method using Differential Evolution (DE). The backfill dimensions, i.e., trench width and height, were assessed, giving the maximum current carrying capacity at an acceptable initial investment cost.

Al-Saud et al. (2007) provided an optimization model for the underground thermal circuit of a power cable based on the generated gradient approach. The authors showed the use of nonlinear optimization combined with thermal finite element field analysis. The sensitivity of cable temperatures to variations in cable circuit parameters is also considered. The proposed finite element thermal analysis was applied to evaluate the cable temperature and its sensitivity with regards to the optimized thermal parameters.

Moutassem and Anders (2010) presented a method for configuring the locations of any number of underground cables to achieve the highest total ampacity. The optimal configuration was determined through a two-level optimization algorithm. At the outer level, a combinatorial optimization based on a Genetic Algorithm (GA) explored different possible configurations which were evaluated according to total ampacity calculated by using a convex optimization algorithm. The convex optimization algorithm that formed the inner level of the optimization procedure was based on the barrier method.

Zarchi et al. (2014) proposed a method to find the optimal cable configuration in a duct assembly considering current harmonics and their effect on sheath losses. The optimal cable position that maximizes the total ampacity was found using PSO and the Shuffled Frog Leaping Algorithm (SFLA).

Del-Pino-López (2014) proposed a shielding cost optimization process that combines finite element models with a genetic algorithm. The dimensions and location of the most cost-effective shield that limits the magnetic field in a specific area without limiting the cable ampacity were found. Both horizontal and inverted-U shielding geometries were tested by applying them to a case study of 9 ($3 \times 3$) ducts in which symmetric three-phase circuits were arranged in three configurations: vertical, horizontal, and triangular. The effects of different materials, phase configurations, shield geometry, and losses were also studied.

The present study proposes a modified Jaya algorithm for the optimization of underground power HV transmission line costs. The Jaya algorithm developed by Rao (2016) can be regarded as a simple and very efficient heuristic optimization technique for solving constrained optimization problems. The main advantage of the Jaya algorithm is that it does not have any algorithm-specific parameters to be tuned, except for the common control parameters of the population size and the number of iterations. The other optimization algorithms like the GA, PSO, SFLA, etc. need the tuning of their respective algorithm-specific parameters in addition to the tuning of the common control parameters. Improper tuning of the algorithm-specific parameters may increase the computational effort or lead to local optima. The modification to the classical Jaya algorithm proposed in this study changes the way of selecting the best and the worst solutions that are used to generate the next population. The performance of the modified Jaya algorithm is compared with the classical Jaya algorithm and the well-known PSO algorithm (Kennedy and Eberhart 1995; Kennedy et al. 2001).

The subject of the study is a 400 kV high-voltage cable line, with XLPE cables arranged in a flat formation and buried 2.0 m beneath the ground level. A thermal backfill layer is included in the system in order to improve heat removal from the cable core. What is important, the material costs (thermal backfill material and cables) are minimized while enhancing the thermal performance of the cable system. The fundamental constraint on the cost function is related to the cable conductor temperature that must not exceed the optimal value for the given cable type. Therefore, a non-linear coupled electric-thermal model of the UPCS is developed and used to determine the maximum temperature within the system. The model is solved using the finite element method with the Author's own codes. The performed optimization designs the thermal bedding cross-section dimensions and selects an appropriate cable conductor area. Based on the unit cost of the thermal backfill and cables, the overall costs of the UPCS installation (excluding excavation works, cable joints, and terminations, as well as additional equipment costs, among others) are determined and minimized using the algorithm proposed in this study. The thermal performance of an advanced thermal backfill material POWERCRETE™ (developed by the Heidelberg Cement Group) is compared with the classical thermal backfills, such as the sand-cement mix and Fluidized Thermal Backfill™ (FTB). To demonstrate the effect of variation in soil thermal conductivity on the performance of different backfills, the performance of thermal backfill materials is studied for the soil thermal conductivity varied from 0.6 to 1.0 W/(m K). Ocłoń et al. (2015a) performed the PSO-based optimization of the UPCS thermal backfill dimensions. However, the study was limited only to minimization of the FTB backfill cross-sectional area for the soil thermal conductivity of 1.0 W/(m K). This chapter presents a more comprehensive optimization approach adopted to minimize the overall material costs of the UPCS installation using a simple yet powerful modified Jaya algorithm. Such a study is important for cable engineers, as it enables them to design the HV or the EHV UPCS in such a way that both thermal performance and material costs are optimized.

The section below presents the details of the UPCS optimization problem.

## 10.1 Optimization Problem

The test case under consideration is the 400 kV underground power cable system arranged in a flat formation shown in Fig. 10.1. The power cables are buried underground in a trench filled with a thermal backfill layer, which improves heat dissipation from the cables. The backfill is situated in homogeneous soil with constant thermal conductivity. The main goal of the optimization is to satisfy the following requirements:

- minimize material unit costs (cables costs + thermal backfill costs per km of the transmission line),

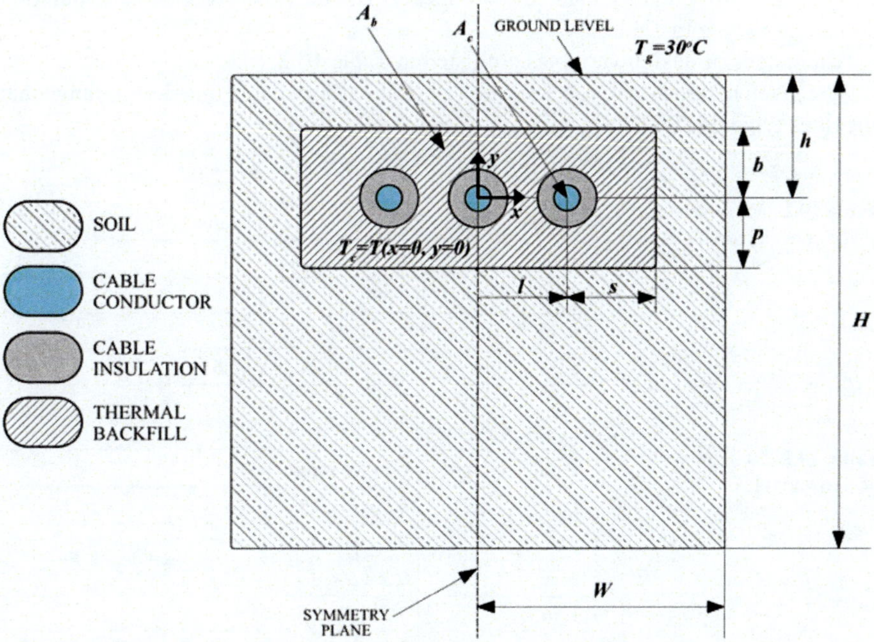


**Fig. 10.1** Underground power cable system with cables arranged in an in-line formation considered for the optimization problem with design variables $l$, $p$, $b$, $s$ and $A_c$

- design a UTL in such a way that the maximum temperature of the cable conductor is equal or close to 65 °C (the optimum temperature of the cable operation as specified by the cable producer).

To simulate the adverse heat transfer condition occurring in the soil during the summer time, the temperature of the ground level is set to 30 °C, as presented in (Popiel et al. 2001).

To minimize the material and installation costs of the UPCS, both thermal backfill ($A_b$) and cable conductor ($A_c$) cross-sectional areas should be kept to a minimum. Also, the transmission line should operate in stable operating conditions with the cable conductor temperature lower or equal to the optimum value specified by the cable producer for a given current loading. It is assumed that the maximum current flowing through the cables $I$ is equal to 1145 A. A significant exceedance in the optimum temperature of the cable conductor may lead to the cable insulation melting and, consequently, to the cable line failure.

The following design variables are considered in the optimization problem (with regard to Fig. 10.1):

$l$—horizontal distance between the cable conductors axes,

$s$—spacing between the right edge of the bedding layer and the side cable axis,

$b$—distance between the conductors axes and the top of the bedding layer,

$p$—distance between the conductors axes and the bottom of the bedding layer,

$A_c$—cable conductor cross-sectional area (selected from the XLPE HV cable design series provided by the cable producer). The dimensions of the computation domain are given in Table 10.1.

The ranges of design variables are listed in Table 10.2:

The given parameters $l, p, b\ s, A_c$ are selected within a cable line design range that specifies typical HV UTL installation cases.

**Table 10.1** Main dimensions of the considered UPCS

| Parameter, unit | value |
|---|---|
| $W$, m | 5 |
| $H$, m | 10 |
| $h$, m | 2 |
| $d_o/d_c$ | 2.67 |

**Table 10.2** Ranges of the design variables

| Parameter, unit | Change range |
|---|---|
| $l$, m | $0.3 < l < 0.5$ |
| $p$, m | $0.2 < p < 0.5$ |
| $b$, m | $0.2 < b < 0.5$ |
| $s$, m | $0.2 < s < 0.5$ |
| $A_c$, m$^2$ | $1{\cdot}10^{-3}$, $1.2{\cdot}10^{-3}$, $1.4{\cdot}10^{-3}$, $1.8{\cdot}10^{-3}$, $2{\cdot}10^{-3}$ |

During the HV UTL installation, the cable cost accounts for a significant part of the overall investment costs. Other investment costs related to the transmission line length rather than to the UPCS cross-sectional area, including excavation costs, cable joints and terminations, as well as additional equipment costs, among others, are not taken into account in the following investigations. The unit costs for a single 400 kV XLPE power cable are listed in Table 10.3. The cable unit costs, given in thousands of euros, are estimated per 1 km of the cable line and rise with an increase in the cable conductor cross-sectional area. Also, the material costs of the thermal backfill are estimated per 1 km of the cable line (with the thermal backfill cross-section of 1.0 m$^2$ and length of 1 km). Therefore, the UPCS installation costs are optimized including both the cables and the backfill material costs. The study includes utilization analysis of the three selected thermal backfill materials that are commonly used by cable engineers during the HV UTL design and installation, such as:

- Sand-Cement Mix with sand-to-Portland cement mass proportion as 12:1;
- Fluidized Thermal Backfill™ (FTB), which is described in (Radhakrishna 1982) as Sand- Gravel-Fly Ash–Cement mix with a given composition and investigated thermal properties,
- POWERCRETE™, a product dedicated to UPCS installation, developed by the Heidelberg Cement Group.

The cable core is assumed to be a braided copper conductor, while the cable insulation is made of cross-linked polyethylene (XLPE) as for most common cases. Figure 10.2 shows the unit cost factor for the three thermal backfill materials under consideration.

**Table 10.3** Unit costs of a single 400 kV XLPE power cable with various conductor cross-sectional areas in comparison with unit costs of different thermal backfill materials (sand-cement mix; FTB and POWERCRETE™) (Ocłoń et al. 2018)

| HV XLPE cable with braided copper conductor | | Sand-Cement Mix | FTB | POWERCRETE™ |
|---|---|---|---|---|
| | | Unit costs (kEuro/km), ($uc_b$) | | |
| | | 20 | 125 | 400* |
| Conductor's cross-sectional area ($A_c$), m$^2$ | Cable unit costs (kEuro / km), ($uc_C$) | Unit Cost factor ($ucf_{SC}$) | Unit Cost factor ($ucf_{FTB}$) | Unit Cost factor ($ucf_{PC}$) |
| 0.001 | 268.8 | 13.44 | 2.1504 | 0.672 |
| 0.0012 | 320.4 | 16.02 | 2.5632 | 0.801 |
| 0.0014 | 373.6 | 18.68 | 2.9888 | 0.934 |
| 0.0016 | 446 | 22.3 | 3.568 | 1.115 |
| 0.0018 | 526.3 | 26.15 | 4.2104 | 1.31575 |
| 0.002 | 626 | 31.3 | 5.008 | 1.565 |

*The author's assumed cost of POWERCRETE™ may differ from the actual costs

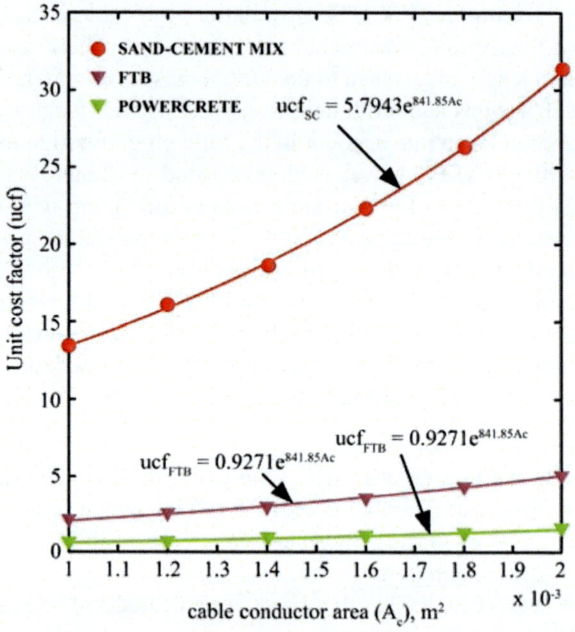


**Fig. 10.2** Unit cost factor for various thermal backfill materials (Sand-Cement Mix, Fluidized Thermal Backfill, and POWERCRETE™) (Ocłoń et al., (2018))

The material costs (costs of backfill layer and power cables) per one km of cable line are given as:

$$C_{total} = 1.5uc_C + A_b uc_b, \tag{10.1}$$

where

$uc_c$—unit costs of single cable per km of the cable line,

$uc_b$—unit costs of thermal backfill material per km of the cable line,

$A_b$—thermal backfill cross-sectional area.

The cable unit costs are multiplied by 1.5 since the UPCS symmetry is assumed, thus, a half of the system (one and a half of the cable) is considered in the computations.

Introducing the unit cost factor $f_{UC}$ as:

$$f_{uc} = \frac{uc_C}{uc_b}, \tag{10.2}$$

and substituting Eqs. (10.2) into (10.1), the following is obtained:

$$C_{total} = 1.5 \cdot ucf \cdot uc_b + A_b uc_b. \tag{10.3}$$

The final cost function, needed to be minimized, is obtained by dividing both sides of expression (10.3) by unit costs of backfill material $uc_b$

$$F_o(A_c, A_b) = \frac{C_{total}}{uc_b} = 1.5ucf + A_b. \tag{10.4}$$

It can be seen that the backfill area is a function of design variables:
$\mathbf{X} = [l, p, b, s, A_c]$:

$$A_b = (l+s)(p+b) - (3/2)\frac{\pi d_o^2}{4}$$
$$= (l+s)(p+b) - 8.3985d_c^2 = (l+s)(p+b) - 10.6933A_c \tag{10.5}$$

where

$d_c$—cable conductor diameter, m

$d_o$—cable outer diameter, m

$A_c$—cable conductor area, m$^2$.

Since the cable outer-to-inner diameter ratio ($d_o/d_c$) is assumed to be equal to 2.67 for the considered XLPE cables series, and the conductor cross-sectional area is given as $A_c = (1/4)\ \pi\ d_c^2$, cost factor $f_{uc}$ can be approximated using the following exponential formula:

$$ucf = ae^{cA_c} \tag{10.6}$$

with constants $a$ and $c$ as given in Table 10.4.

Consequently, substituting Eqs. (10.6) in (10.5), the final form of the cost function is obtained.

$$F_o(l, p, b, s, A_c) = \frac{C_{total}}{uc_B} = ae^{cA_c} + (l+s)(p+b) - 10.6933A_c, \tag{10.7}$$

XLPE cables are one of the commonly used in the UTL. Since, the XLPE cable insulation is made of cross-linked polyethylene, the optimum conductor temperature for extra-high voltage (EHV) cables (according to cable producers) is equal to $T_{opt}$ = 65 °C. The optimum temperature ensures safe operation of the cable significantly below the perilous range of temperatures below 90 °C, which is the polyethylene melting point. Therefore, exceeding the $T_{opt}$ temperature value should eliminate the set of the design variables from the solution domain. In such a case, the static penalty function defined by Eq. (10.8) is employed:

**Table 10.4** Coefficients used in the unit cost factor model for the considered different thermal backfill materials

| Backfill material | *a* | *c* |
|---|---|---|
| Sand cement mix | 5.7943 | 841.85 |
| FTB | 0.9271 | |
| POWERCRETE™ | 0.2897 | |

$$PF = 10 \cdot \left(T_c(l, p, b, s, A_b) - T_{opt}\right). \tag{10.8}$$

The penalty function *PF* is added to Eq. (10.7) and applied when the maximum temperature of the central cable conductor $T_c = T(x = 0, y = 0)$ is greater than $T_{opt}$.

$$\begin{cases} F_o(l, p, b, s, A_c) = ae^{cA_c} + (l + s)(p + b) - 10.6933A_c + PF, \text{ for } T_c - T_{opt} > 0 \\ F_o(l, p, b, s, A_c) = ae^{cA_c} + (l + s)(p + b) - 10.6933A_c, \text{ for } T_c - T_{opt} \le 0 \end{cases} \tag{10.9}$$

The following section presents the details of the determination of the cable conductor temperature.

## 10.2 Electric-thermal Model of the Considered UTL

In order to determine the temperature distribution within the UPCS, its thermal model is developed. The value of the maximum temperature of the cable conductor is needed to calculate the penalty function *PF* in Eq. (10.8). The heat transfer phenomenon within the UPCS can be considered as a steady-state and two-dimensional case. It is assumed that the current *I* rating is equal to 1145 A (the value of optimal current loading corresponding to temperature $T_{opt}$) and the loading will not exceed this value during the cable line operation. Also, it is assumed that the loading is the same for each cable in the system. A half of the physical model, as shown in Fig. 10.3, is used within the computational domain since the temperature distribution pattern is expected to provide mirror symmetry with respect to the symmetry plane.

The UPCS computational domain consists of four subdomains:

- soil domain (*s*),
- thermal backfill domain (*b*),
- copper cable conductor domain (*c*),
- cable insulation domain (*ins*).

The heat conduction equation (Eq. (10.10)) is solved to determine the temperature distribution within the considered UPCS (Ocłoń et al. 2015b):

$$\frac{\partial}{\partial x}\left[k\frac{\partial T(x, y)}{\partial x}\right] + \frac{\partial}{\partial y}\left[k\frac{\partial T(x, y)}{\partial y}\right] = -q_v(T(x, y)), \tag{10.10}$$

where

*x, y*—Cartesian coordinates of the selected point belonging to the computational domain,

*k*—thermal conductivity, W/(m K),

$q_v(T(x, y))$—temperature-dependent heat source (applied only to the domain of the cable conductor), W/m$^3$.

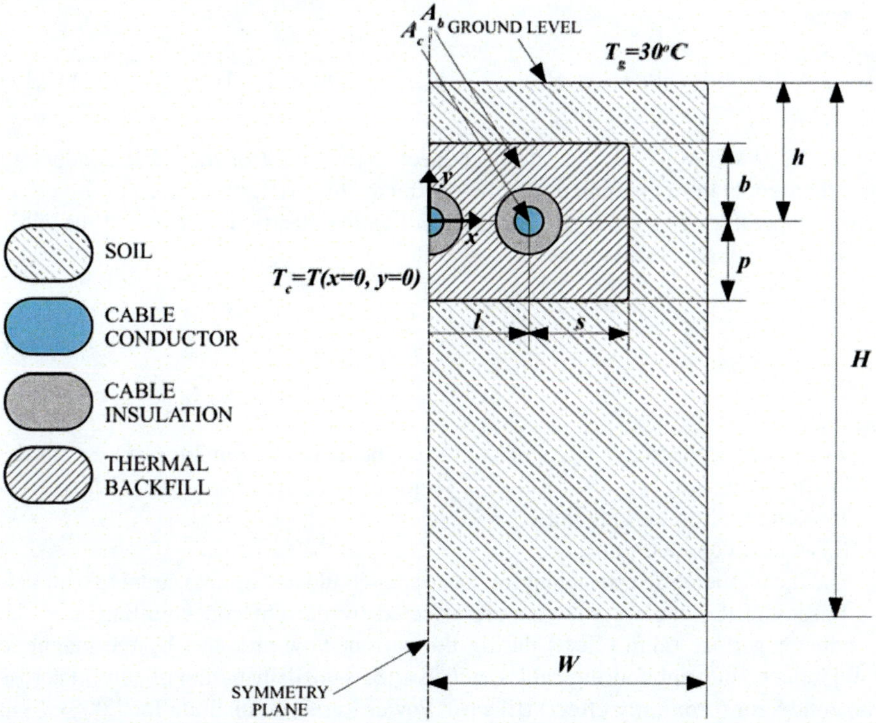


**Fig. 10.3** Half of the UPCS domain considered in the computation of the cable conductor temperature $T_c$

Considering the heat conduction equation (Eq. (10.10)), the heat source term is defined as:

$$q_v(T(x, y)) = \frac{\Delta Q(T(x, y), l, A_c)}{A_c}. \tag{10.11}$$

Joule's law expresses the power cable heat losses by introducing current rating *I* and alternating current (AC) electric resistance Re,AC to the equation:

$$\Delta Q = I^2 R_{e,AC}(T(x, y), l, A_c) \tag{10.12}$$

The cable conductor resistance exhibits a greater electric resistance in the case of the alternating current flow ($R_{e,AC}$) than for direct current ($R_{e,DC}$). Skin and proximity effects, which are the main causes of the observed changes, may be introduced to Eq. (10.12) by using $y_s$ and $y_p$ coefficients, respectively:

$$R_{e,AC} = R_{e,DC}(1 + y_s + y_p), \tag{10.13}$$

with

$$R_{e,DC} = R_{e,ref}\left[1 + \alpha_{ref}(T(x, y) - T_{ref})\right] \tag{10.14}$$

where $R_{e,ref}$ and $\alpha_{e,ref}$ are the reference electric resistance of the cable conductors and the temperature coefficient for the conductor material, both given at the reference temperature $T_{ref} = 20$ °C. For Eq. (10.14), the electric resistance of the cable conductor may be calculated as:

$$R_{e,ref} = \frac{\rho_{20} C}{A_c} \tag{10.15}$$

where

$\rho_{20}$—specific electrical resistance of the copper conductor in 20 °C, Ω ·m; for the following case study $\rho_{20} = 1.7241\ 10^{-8}$ Ω ·m,

$C$—length of the sample, m; $C = 1.0$ m,

$A_c$—cable conductor area, m$^2$,

and the temperature coefficient for the copper conductor $\alpha_{e,ref}$ is equal to 0.00393.

Losses in the cable conductor are affected by current concentrating near the conductor surface (skin effect) during the current flow and also by the magnetic field induced in neighbouring cables, which affects the distribution of current across the conductor (proximity effect) (Electric Power Research Institute Inc 2003). Skin and proximity effect coefficients $y_s$ and $y_p$ are calculated respectively as (Hiivala and Landinger 2014):

$$y_s = \frac{x_s^2}{192 + 0.8x_s^2},\ x_s = \frac{8\pi f}{R_{e,DC}} \cdot 10^{-7} k_s \tag{10.16}$$

$$y_p = \frac{x_p^2}{192 + 0.8x_p^2}\left(\frac{d_c}{l}\right)^2\left[0.312\left(\frac{d_c}{l}\right)^2 + \frac{1.18}{\frac{x_p^2}{192+0.8x_p^2} + 0.27}\right],$$

$$x_p = \frac{8\pi f}{R_{e,DC}} \cdot 10^{-7} k_p \tag{10.17}$$

where $k_s$ and $k_p$ are skin and proximity effect correction factors equal to 0.435 and 0.37, respectively, for the case of the segmented conductor type [. In Eqs. (10.16) and (10.17), $f = 50$ Hz is an alternating current frequency and $l$ is the distance between neighbouring conductor axes.

The real system of the underground electrical energy transmission line consists of three power cables. The cables are located at depth $H = 2.0$ m below the ground level. However, due to the expected mirror symmetry of the temperature field solution pattern, only a half of the real system is modelled. The square domain with the height of 10.0 m and width of 5.0 m is used in the computations (Fig. 10.1). It is assumed that the right, left and bottom edges of the boundary region are thermally insulated (i.e. no heat flow exists in the direction perpendicular to the domain edge). At the top edge (the ground level), temperature $T_g$ is set to 30 °C to simulate the adverse heat transfer condition in the summer period. Equation (10.18) summarizes the types of boundary conditions applied when solving Eq. (10.10):

$$\begin{aligned}
&k\frac{\partial T}{\partial x}\bigg|_{x=W} = 0, \text{ for the boundary region right edge,}\\
&k\frac{\partial T}{\partial x}\bigg|_{x=0} = 0, \text{ for the symmetry plane,}\\
&k\frac{\partial T}{\partial y}\bigg|_{y=H-h} = 0, \text{ for the boundary region bottom edge,}\\
&T(x, y = h) = 30^{\circ}\text{C, for the boundary region top edge.}
\end{aligned} \tag{10.18}$$

The heat conduction equation describes the mathematical model of heat transfer processes (Eq. 10.10) with a non-linear (temperature-dependent) source term Eq. (10.11). The boundary conditions for Eq. (10.10) are given by Eq. (10.18). Equations (10.12–10.17) give the calculation method for determining the source term. The heat conduction equation Eq. (10.1) is discretized using the Finite Element Method (FEM) and solved employing the Jacobi iterative scheme. The solution here is the nodal temperature distribution in the entire cable system. The maximum value of the determined temperature $T_c = T\ (x = 0, y = 0)$ is considered in determining the value of cost function $F_o$ in Eq. (10.9).

The algorithm for FEM simulations of the UPCS was presented by the author in a previous study (Ocłoń et al. 2015b). In Appendix A of that paper, a detailed algorithm of FEM simulations was given. Also, the FEM codes (written in MATLAB) for the UPCS analysis were attached. The codes were listed in the supplementary material section. The present study uses the codes developed in (Ocłoń et al. 2015b) but the emphasis is not on the FEM simulation but on the optimization of the UPCS. Other useful and validated approaches to simulations of the thermal and electrical performance of the UPCS are available in the literature (Kroener et al. 2014; Papagiannopoulos et al. 2013; Salata et al. 2015; Lieto Vollaro and Vallati 2013; Lieto Vollaro et al. 2011a,2011b; Galli and Vallati 2011; Hwang and Jiang 2003). The proposed UPCS model, based on the FEM method, is a compromise between computational complexity and accuracy. The model performance is high since the efficient method presented by Cuvelier et al. (2013) is used to determine finite element matrices, and the FEM matrices assembly is done in a short period of time. Also, the temperature field is evaluated by solving the two-dimensional heat conduction equation within the entire cross-section across the cable surrounding the backfill and the soil in the computational domain. The proposed model is simplified and consider only heat losses in power cable conductor. To consider the heat loss in cable insulation, the readers may refer to the author's works (Ocłoń et al. 2021) and (Ocłoń et al. 2018).

## 10.3 Thermal Conductivity of the Computational Subdomains

To solve the heat conduction equation, the thermal conductivity of each subdomain within the discrete model should be given. Therefore, the materials of each component of the underground power cable system with the respective thermal conductivities are provided.

### *10.3.1 Thermal Conductivity of Cable Layers*

In this section, a system of three high-voltage 400 kV XLPE cables buried in the ground in a flat (in-line) arrangement is analysed. The considered cables consist of a solid cable core, cross-linked polyethylene (XLPE) insulation, a metallic sheath, and non-metallic outer covering. Additionally, each cable layer is separated using an insulating tape. The copper cable conductor is additionally segmented to reduce current losses (TELE-FONIKA, 2012). The layout and arrangement of the layers in the considered power cable are shown in Fig. 10.4.

Usually, high-voltage power cables consist of four main layers: the cable conductor, insulation, sheath and jacket, as shown in Fig. 10.5.

Table 10.5 lists the materials and thermal conductivities of individual layers of the cable.

In the present study, the simplified model of the 400 kV XLPE cable is considered. The model consists of only two layers, i.e. the cable conductor and the equivalent cable insulation. The model makes it possible to reduce the number of finite elements in a discrete model of the UPCS. It is assumed that the cable insulation, sheath, and jacket layers are replaced with equivalent insulation layer $d_O$ with constant thermal conductivity $k_{eqv}$ given by Eq. (10.19):

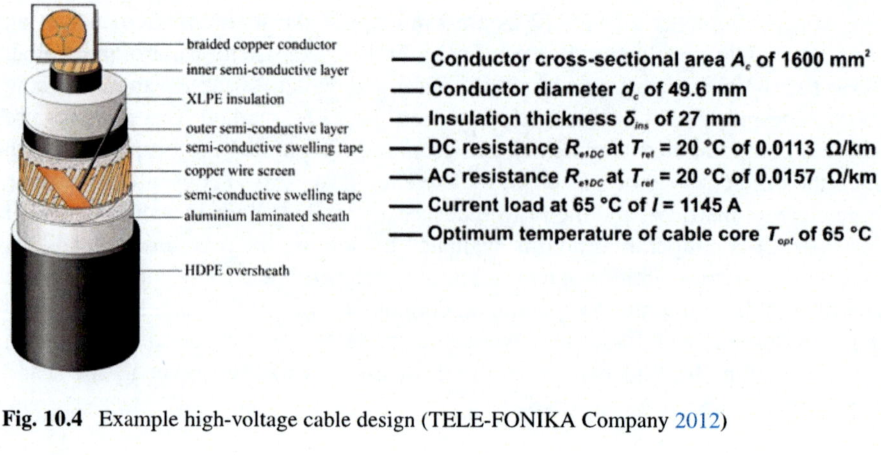


**Fig. 10.4** Example high-voltage cable design (TELE-FONIKA Company 2012)

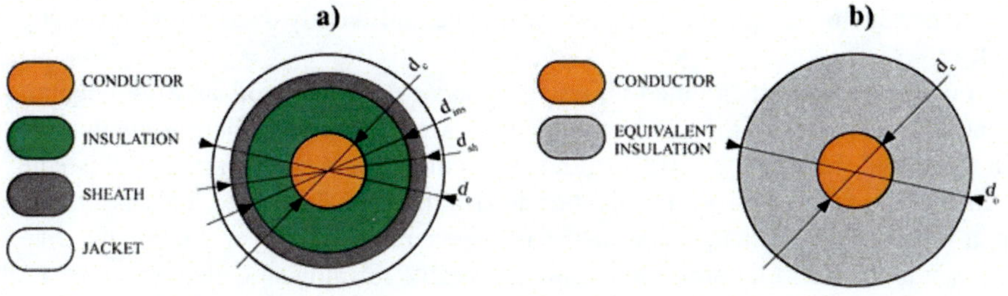


**Fig. 10.5** **a** Detailed and **b** simplified model of a 400 kV power cable.

**Table 10.5** Thermal properties and outer diameters of the power cable layout materials (referred to Fig. 10.4a)

| Cable layer | Material | Diameter, mm | Thermal conductivity, W/(m·K) |
|---|---|---|---|
| Conductor | Copper | $d_c = 49.6$ | 400 |
| Insulation | XLPE | $d_{ins} = 110.6$ | 0.2875 |
| Sheath | Copper | $d_{sh} = 123.0$ | 400 |
| Jacket | HDPE | $d_o = 133.6$ | 0.2875 |
| Equivalent insulation | | $d_o = 133.6$ | 0.3232 |

$$k_{eqv} = \frac{\ln\left|\frac{d_o}{d_c}\right|}{\frac{\ln\left|\frac{d_{ins}}{d_c}\right|}{k_{ins}} + \frac{\ln\left|\frac{d_{sh}}{d_{ins}}\right|}{k_{sh}} + \frac{\ln\left|\frac{d_o}{d_{sh}}\right|}{k_{ja}}} \tag{10.19}$$

where $k_{ins}$, $k_{sh}$, and $k_{ja}$ represent thermal conductivities of the cable layers, insulation, sheath, and jacket, respectively.

### *10.3.2 Thermal Conductivity of the Soil and Backfill Materials*

IEC standards consider the soil thermal conductivity as equal to 1.0 W/(m·K) with respect to the region of Poland (IEC 1995 ). Kroener et al. (2014) showed that the soil thermal conductivity depends on various parameters, including e.g. bulk density, water content, clay content and temperature. The comprehensive Campbell-de Vries model (Campbell et al. 1994) of soil thermal conductivity incorporates all those parameters. In regions with the soil low water content, the soil thermal conductivity can decrease to 0.3 W/(m·K), or even to lower values. Typically, the decrease in the soil thermal conductivity is five- to tenfold when the soil moisture content drops to zero (Campbell et al. 1994). Therefore, the assumption that the soil thermal conductivity is equal to 1.0 W/(m·K) may lead to the underestimation of the cable conductor temperature in cases when the soil locally dries out (Gouda et al. 2011).

For the reasons mentioned above, this paper considers soil as a homogenous material and performs the UPCS optimization for the soil mean thermal conductivity in the range from 0.6 to 1.0 W/(m·K). The computations performed in this study consider the thermal backfill properties determined in the dry state. This assumption is a result of the fact that due to the heat generated by power cables, the water present in the backfill pores may evaporate locally, forming so-called dry zones (An EU Strategy on Heating and Cooling 2016). Therefore, to assure the safety factor in designing UTLs, and to protect the UTL cables from overheating, the system is designed for the lowest possible thermal conductivity of the thermal bedding layer. This design approach makes it possible to ensure a safety margin because a higher thermal conductivity of the thermal backfill is expected in operating conditions.

The thermal conductivity of dry backfill is considerably higher compared to dry soil. Sand-Cement Mix has been used in Europe as a backfill material in the cables burying technology for decades. The typical material in use is the mix with the 12:1 sand-cement ratio (Adams and Baljet 1968). This type of thermal backfill is inexpensive (~20 Euros per one cubic meter). However, its thermal conductivity in the dry state is equal to 1.0 W/(m K) or slightly below this value. Another popular material is Fluidized Thermal Backfill (sand and Portland cement mix in proportions of 10:1 with fly ash acting as a fluidizer, and other additives). The FTB backfill is commonly used in the USA. The best available FTB material is the SGFC (a mix of sand, gravel, fly ash and cement), ensuring the thermal conductivity of 1.54 W/(m K) in the dry state (Radhakrishna 1982). The FTB unit costs are approximately 125 Euro per one cubic meter (Davis et al. 2010). Recently, the Heidelberg Cement Group developed a thermal backfill material named POWERCRETE™ with the thermal conductivity of up to 3.0 W/(m K) in the dry state. This value is quite high and improves heat dissipation from buried cables significantly, especially in the vicinity of the cable line. The costs of POWERCRETE™ are approximated as 400 Euro per one cubic meter. The real costs may differ from the Author's assumption. Table 10.6 shows the thermal conductivities of the considered thermal backfill materials.

This paper attempts to find cost-effective solutions for the UPCS design with respect to the condition that the cable conductor temperature does not exceed the optimum temperature of the cable operation, i.e. 65 °C with the considered current rating of 1145 A. The design solutions are found for the soil thermal conductivity

**Table 10.6** Thermal conductivity of the thermal backfill materials (Sand-Cement Mix; FTB and POWERCRETE™) and soil considered in the computations

| Material | Thermal conductivity, W/(m K) |
|---|---|
| Sand-Cement Mix (with proportions of 1:12) | 1.0 |
| FTB | 1.54 |
| POWERCRETE™ | 3.0 |
| Soil | 0.6–1.0 |

varied from 0.6 to 1.0 W/(m K). Moreover, the cable line costs are optimized using the modified version of a recently proposed Jaya optimization algorithm.

## 10.4 Optimization Algorithm

The following section presents the Jaya algorithm and its modified version used for the UPCS optimization.

### *10.4.1 Jaya Algorithm*

The Jaya algorithm was developed by Rao (2016). Let $F(X)$ be the objective function to be minimized (or maximized). At any iteration *i*, assume that there are '*m*' design variables (i.e. $j = 1,2,\ldots,m$) and '*n*' candidate solutions (i.e. population size $k = 1,2,\ldots,n$). Let the best candidate *best* achieve the best value of $f(x)$ (i.e. $f(x)_{best}$) in the entire set of candidate solutions, and the worst candidate *worst*—the worst value of $f(x)$ (i.e. $f(x)_{worst}$) in the entire set of candidate solutions. If $X_{j,k,i}$ is the value of the *j*th variable for the *k*th candidate during the *i*th iteration, then this value is modified as per the following Eq. (10.20):

$$X'_{j,k,i} = X_{j,k,i} + r_{1,j,i}\left(X_{j,best,i} - \left|X_{j,k,i}\right|\right) - r_{2,j,i}\left(X_{j,worst,i} - \left|X_{j,k,i}\right|\right) \quad (10.20)$$

where $X_{j,best,i}$ is the value of variable *j* for the best candidate and $X_{j,worst,i}$ is the value of variable *j* for the worst candidate. $X'_{j,k,i}$ is the updated value of $X_{j,k,i}$ and $r_{1,j,i}$ and $r_{2,j,i}$ are two random numbers for the *j*th variable during the *i*th iteration from the interval [0,1]. The term "$r_{1,j,i}\left(X_{j,best,i} - \left|X_{j,k,i}\right|\right)$" indicates the tendency of the solution to move closer to the best solution and the term "$-r_{2,j,i}\left(X_{j,worst,i} - \left|X_{j,k,i}\right|\right)$" indicates the solution tendency to avoid the worst solution. $X'_{j,k,i}$ is accepted if it gives a better function value. All the accepted function values at the end of a given iteration are maintained and these values become the input to the next iteration. Figure 10.6 shows the flowchart of the Jaya algorithm.

The algorithm always tries to get closer to success (i.e. achieve the best solution) and avoid failure (i.e. move away from the worst solution). The algorithm strives to become victorious by achieving the best solution and hence it is named as *Jaya* (a Sanskrit word meaning *victory*). Rao (2016) proved that the Jaya algorithm performs better than many other optimization algorithms for solving constrained benchmark problems.

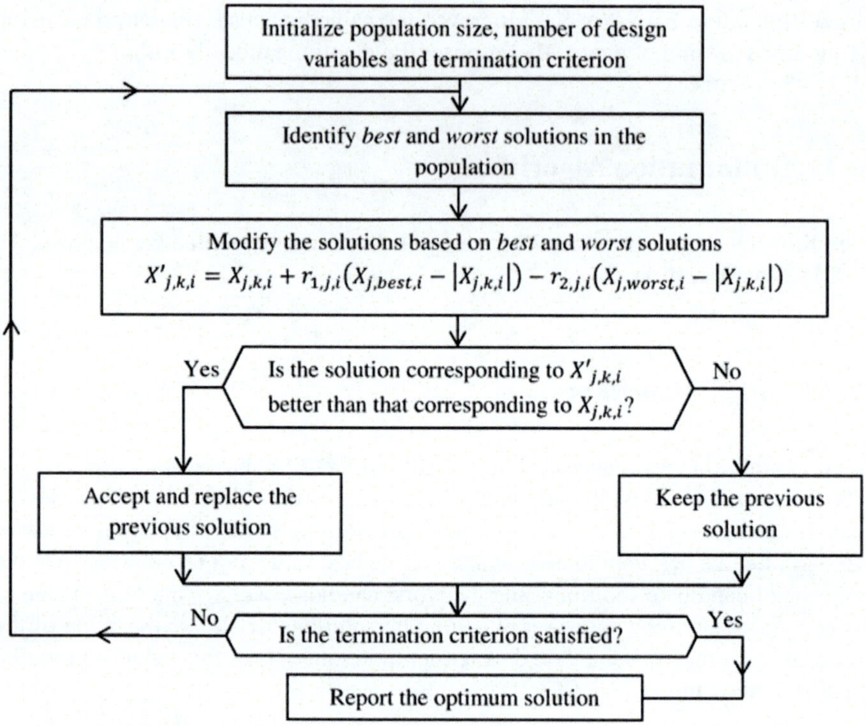


**Fig. 10.6** Flowchart of the Jaya algorithm (Rao 2016)

### *10.4.2 Modified Jaya Algorithm*

The modified Jaya algorithm is described by Oclon et al. (2018). The following modification to the classical Jaya algorithm is proposed to improve the search for the optimum solution. The expression

$$r_{1,j,i}\left(X_{j,best,i} - \left|X_{j,k,i}\right|\right) - r_{2,j,i}\left(X_{j,worst,i} - \left|X_{j,k,i}\right|\right)$$

in Eq. (10.20) is replaced with:

$$r_{1,j,i}\left(B[rb] - \left|X_{j,k,i}\right|\right) - r_{2,j,i}\left(W[rw] - \left|X_{j,k,i}\right|\right)$$

Where, $B$ is a subset of three best solutions (with the lowest cost function value) and $W$ is a subset of three worst solutions (with the highest cost function value).

The algorithm generates two integer random numbers *rb*, *rw*—1, 2 or 3 for each individual candidate and the final form of the candidate's position is updated as:

$$X'_{j,k,i} = X_{j,k,i} + r_{1,j,i}\left(B[rb] - \left|X_{j,k,i}\right|\right) - r_{2,j,i}\left(W[rw] - \left|X_{j,k,i}\right|\right) \quad (10.21)$$

The modified Jaya algorithm was tested with different population sizes for the design of the UPCS shown in Fig. 10.1.

## 10.5 Results and Discussion

This section presents the results of the tests. Also, to verify the results of the modified Jaya algorithm, the classical Jaya algorithm and the Particle Swarm Optimization (PSO) solver available in MATLAB2015b are also employed.

### 10.5.1 Performance Analysis of the Modified Jaya Algorithm

The performance tests of the modified Jaya algorithm were conducted for the data given in Table 10.7.

The convergence graphs for design variables *l*, *p*, *b* and *s* are shown in Fig. 10.7. Figure 10.8 presents the convergence graph for $A_c$. The variations in the algorithm output parameters (cost function $F_o$, thermal backfill area $A_b$ and the maximum temperature of the cable conductor $T_c$) are also shown in Fig. 10.8. The population

**Table 10.7** Test case data for the performance analysis of the modified Jaya algorithm

| $k_s$, W/ (m·K) | $k_b$, W/(m·K) | *I*, A | *n* | Max iter |
|---|---|---|---|---|
| 0.7 | 1.57 | 1.145 | 10;15;20;25;30;35;40 | 100 |

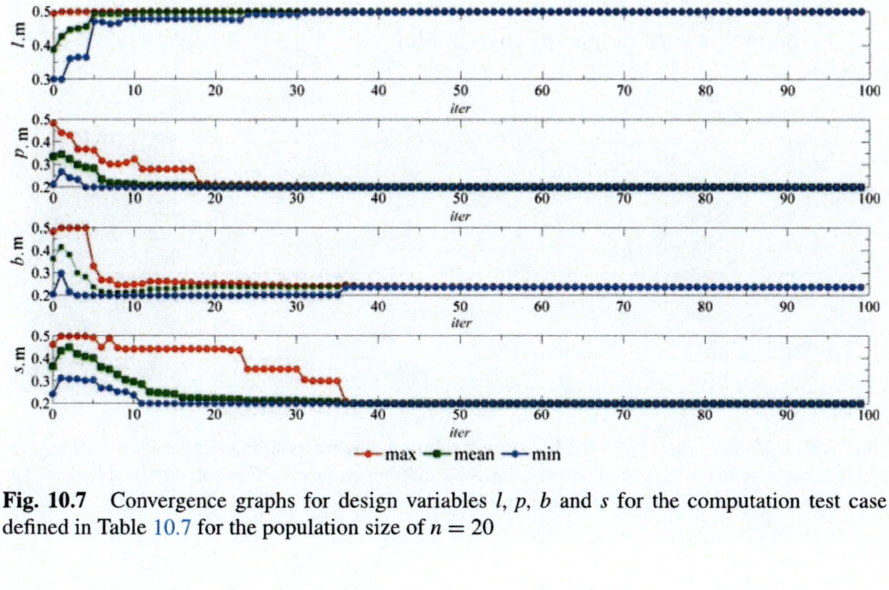


**Fig. 10.7** Convergence graphs for design variables *l*, *p*, *b* and *s* for the computation test case defined in Table 10.7 for the population size of $n = 20$

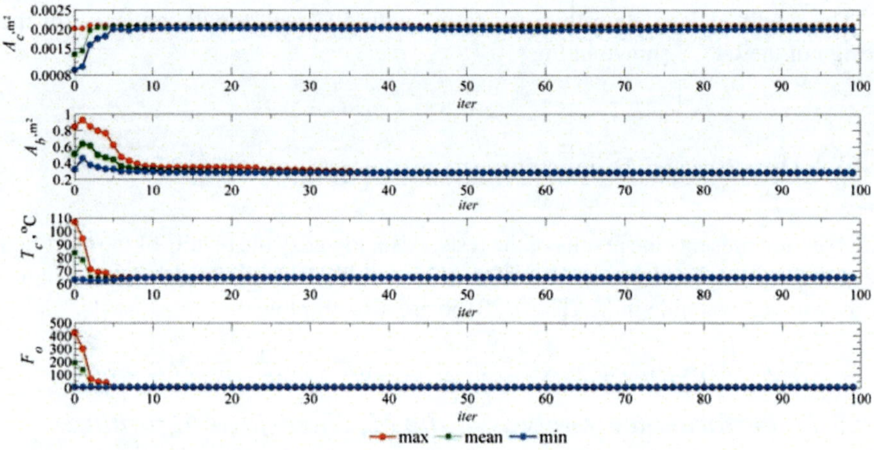


**Fig. 10.8** Convergence graphs for design variable $A_c$ and output parameters $F_o$, $T_c$, and $A_b$ for the computation test case defined in Table 10.7 for the population size of $n = 20$

size used for the presented computational results is $n = 20$. Figures 10.7 and 10.8 indicate that the modified Jaya algorithm converges within less than 40 iterations for all the design variables and output parameters. Analysing Fig. 10.9, it can be seen that the temperature of the cable conductor converged to the desired value of 65 ºC. The minimum, maximum and mean values of the design variables ($l$, $p$, $b$, $s$ and $A_c$) and output parameters ($T_c$, $F_o$, $A_b$) remain the same after the solution convergence is achieved. Two primary advantages of the modified Jaya algorithm can be observed:

1. The algorithm converges very fast (in less than half of the maximum number of iterations under consideration).
2. There is no need to adjust any parameters in the algorithm in Eq. (10.21), and therefore it is very simple in programming.

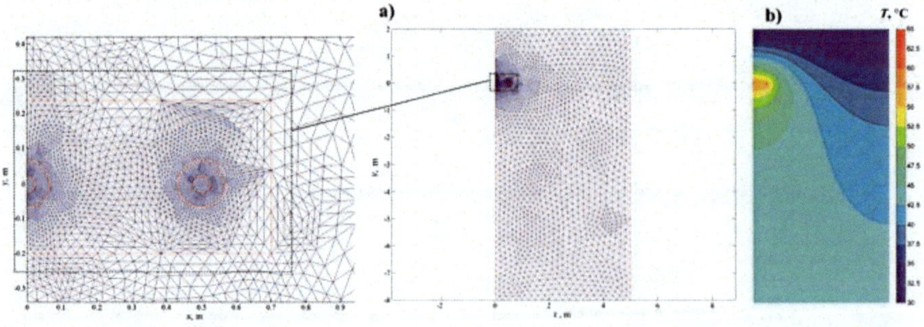


**Fig. 10.9** **a** Mesh generated for the best-found solution, **b** corresponding temperature distribution within the analysed underground power cable system; temperature distribution obtained for the test case defined in Table 10.7 for the population size of $n = 20$.

The geometry and mesh of the UPCS obtained after the design optimization are shown in Fig. 10.9a. Figure 10.9b shows the obtained temperature distribution within the underground power cable system. It should be noted that the maximum temperature of the cable conductor is equal to 65 ºC, as the value expected in the search for a feasible solution.

Further tests were performed to validate the results of the modified Jaya algorithm. As mentioned in the previous section, the optimization results are compared with the classical version of the Jaya algorithm and with the Particle Swarm Optimization (PSO) solver available in MATLAB2015a. The default MATLAB settings are employed for the PSO algorithm during the test computation.

Table 10.8 shows the best-found solution for the design variables (*l, p, b, s* and $A_c$); the minimum and mean values of the cost function ($F_{o,min}$ and $F_{o,mean}$) as well as the maximum temperature of the cable conductor $T_c$ and the corresponding cross-sectional area of the thermal backfill layer $A_b$. The computations were performed for different population sizes *n* varied from 10 to 40. Table 10.8 indicates that the proposed modified Jaya algorithm performs better than the classical Jaya algorithm for the optimization problem defined in this study.

**Table 10.8** Comparison of the best-found solution: design variables (l, p, b, s, $A_c$) and output parameters ($A_b$, $T_c$, $F_o$,min and $F_{o,mean}$) obtained for the modified Jaya, classical Jaya and PSO algorithms for population sizes varied from 10 to 40

| Modified Java | | | | | | | | | |
|---|---|---|---|---|---|---|---|---|---|
| *n* | *l,m* | *p, m* | *b, m* | *s, m* | $A_c$ $10^{-3}$ $m^2$ | $A_b$, $m^2$ | $F_o$, *min* | $T_c$, °*C* | $F_{o,mean}$ |
| 10 | 0.5 | 0.2 | 0.23811 | 0.2 | 2 | 0.28531 | 7.77458 | 64.99 | 7.77460 |
| 20 | 0.5 | 0.2 | 0.23808 | 0.2 | 2 | 0.28527 | 7.77454 | 64.99 | 7.77460 |
| 30 | 0.5 | 0.2 | 0.23812 | 0.2 | 2 | 0.28530 | 7.77457 | 64.99 | 7.77462 |
| 40 | 0.5 | 0.2 | 0.23818 | 0.2 | 2 | 0.28534 | 7.77461 | 65.00 | 7.77463 |
| *Jaya* | | | | | | | | | |
| *n* | *l,m* | *p, m* | *b, m* | *s, m* | $A_c$ $10^{-3}$ $m^2$ | $A_b$, $m^2$ | $F_o$, *min* | $T_c$, °*C* | $F_{o,mean}$ |
| 10 | 0.44 | 0.2 | 0.37280 | 0.2 | 2 | 0.34293 | 7.83220 | 64.99 | 7.83220 |
| 20 | 0.5 | 0.2 | 0.20001 | 0.3 | 2 | 0.29905 | 7.78832 | 64.99 | 7.78831 |
| 30 | 0.5 | 0.2 | 0.23819 | 0.2 | 2 | 0.28535 | 7.77462 | 64.99 | 7.77462 |
| 40 | 0.5 | 0.2 | 0.23808 | 0.2 | 2 | 0.28530 | 7.77457 | 65.00 | 7.77461 |
| *PSO* | | | | | | | | | |
| *n* | *l,m* | *p, m* | *b, m* | *s, m* | $A_c$ $10^{-3}$ $m^2$ | $A_b$, $m^2$ | $F_o$, *min* | $T_c$, °*C* | $F_{o,mean}$ |
| 10 | 0.5 | 0.2 | 0.23409 | 0.2 | 2 | 0.28638 | 7.77565 | 65.00 | 7.78954 |
| 20 | 0.5 | 0.2 | 0.23817 | 0.2 | 2 | 0.28533 | 7.77460 | 65.00 | 7.79163 |
| 30 | 0.5 | 0.2 | 0.23810 | 0.2 | 2 | 0.28528 | 7.77455 | 65.00 | 7.79310 |
| 40 | 0.5 | 0.2 | 0.23816 | 0.2 | 2 | 0.28532 | 7.77459 | 65.00 | 7.79574 |

The best-found solution remains practically the same for all studied population sizes for the modified Jaya algorithm. Also, compared to the PSO optimization solver, the performance of the modified Jaya algorithm is the same or better for individual cases. The mean value of the cost function obtained from the modified Jaya algorithm is lower than that given by the PSO algorithm in all cases.

### 10.5.2 Material Cost Optimization and Thermal Performance Analysis for a 400 kV ULT

The benchmark problem performed in the previous section proved the better performance of the modified Jaya algorithm. In the present section the modified Jaya algorithm, with the population size of $n = 20$ is used to find the best possible design of the UPCS (shown in Fig. 10.1 and the geometry given in Tables 10.1 and 10.2). The population size of $n = 20$ is selected because the *minimum* value of the cost function $F_{o,min}$ is the lowest for this population size value in Table 10.8.

The optimization of the UPCS design is performed for the soil thermal conductivity varied from 0.6 to 1.0 W/(m·K) and three thermal backfill types (sand-cement mix, FTB and POWERCRETE™). Since the optimization problem is formulated for a half of the considered UPCS, the values of total costs, defined by Eq. (10.1), and the cable backfill area $A_b$ obtained from the optimization solver, are doubled and presented in the respective figures.

Fig. 10.9a shows the variation of the UPCS costs for different thermal backfill materials and various soil thermal conductivity values. It can be observed from Fig. 10.10a that only the POWERCRETE™ backfill (PC) makes it possible to achieve the optimum solution for the worst scenario, i.e. the soil thermal conductivity of 0.6 W/(m·K). In this case, it is possible to maintain the cable conductor temperature of 65 °C (Fig. 10.10b) for the thermal backfill area of 0.92 $m^2$ and the cable conductor area of 2000 $mm^2$. For the soil thermal conductivity of 0.6 W/(m·K), it can be seen that the commonly used sand-cement mix (SCM) performs poorly. The thermal backfill cross-sectional area of 2 $m^2$ (Fig. 10.10d) and the largest possible cable cross-sectional area of 2000 $mm^2$ (Fig. 10.10c) made it possible to obtain the cable conductor temperature of 68.5 °C. This value exceeds the optimum temperature $T_{opt} = 65$ °C (cf. Figure 10.10b).

The cable conductor maximum temperature, equal to 65 °C (Fig. 10.10b), is observed for the UPCS model when the thermal backfill is made using FTB. Therefore, the FTB backfill also does not guarantee proper performance under the soil thermal conductivity of 0.6 W/(m·K).

The application of the sand-cement mix (SCM) as thermal backfill for the considered UPCS does not result in a satisfying solution if the soil thermal conductivity is less than 0.75 W/(m·K). In such cases, the cable conductor temperature exceeds the optimum temperature of $T_{opt} = 65$ °C. Therefore, the obtained solution is not considered as feasible (Fig. 10.10a). The application of the FTB backfill resulted

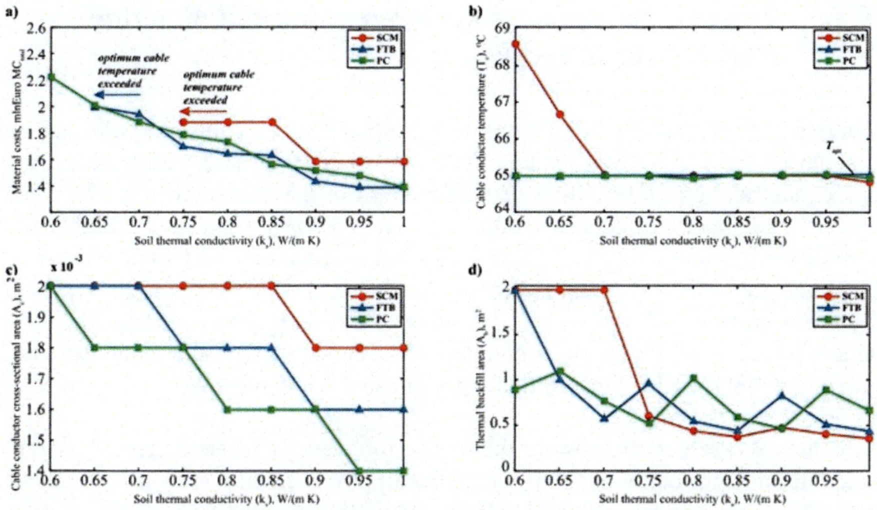


**Fig. 10.10** The best-found solutions obtained using the modified Jaya algorithm for overall material costs of the UPCS and the corresponding: cable conductor maximum temperature, cable conductor cross-sectional area and thermal backfill cross-sectional area

in an appropriate solution for the soil thermal conductivity in the range of 0.65–1.0 W/(m K). Figure 10.10a indicates that the total material costs are the highest for the SCM backfill because larger cable conductor areas are applied. For the analysed system, and for the soil thermal conductivity in the range of 0.75–0.95 W/(m·K), the application of the FTB layer is the most cost-effective option of all the thermal backfill materials under consideration. For the case with the soil thermal conductivity of 1.0 W/(m·K), as suggested by the IEC standards (IEC(1995)), the FTB and the POWERCRETE™ backfills produce similar overall material costs. POWERCRETE™ makes it possible to reduce the cable conductor cross-sectional area from 1,600 to 1,400 mm$^2$ (Fig. 10.10c). However, the backfill area is 40% larger compared to the FTB backfill (Fig. 10.10d). Therefore, the overall material costs remain at the same level.

The main conclusion that can be drawn from Fig. 10.10 is that the application of modern thermal backfills (FTB or POWERCRETE™) instead of the cheaper sand-cement mix can decrease overall material costs of the UPCS while the cable conductor temperature is equal to its optimum value. The modern thermal backfills improve the UPCS thermal performance considerably compared to the sand-cement mix, which is frequently used by cable engineers due to the lower cost. The reduction in the UPCS overall cost is due to the decrease in the cable conductor cross-sectional area. Because unit costs of 400 kV XLPE cables are significantly higher than the costs of backfill materials (Table 10.3), the UPCS overall costs decrease.

### 10.5.3 Thermal Performance of Different Backfill Materials Under Variable Loading

Electrical energy producers need to ensure the immunity of the cable line performance to slight changes in current loading (Thue 2011). The variations may be caused by the electrical grid instabilities or malfunctions of the grid segments. The cable conductor temperature increases with the current loading because the higher the current, the larger the amount of heat generated in the cable conductor. Therefore, the primary aim of the thermal backfill application is to ensure that a slight variation in current loading will not cause overheating of the cable insulation material. The cable producer specifies the maximum allowable temperature of the cable operation and for the 400 kV XLPE underground power cable it is equal to 90 °C (Ocłoń et al. 2015c).

Figures 10.11a, b and c show the thermal performance of different thermal backfill materials under the current loading varied from 1,000 to 2,000 A and the soil thermal conductivity ranging from 0.6 to 1.0 W/(m·K).

Table 10.9 lists the values of design variables that define the thermal backfill layer considered in the computations.

The set of the values presented in Table 10.9 is applied to the model of the cable line. The values are selected based on the design data of the 400 kV cable line which is planned to be installed in Poland. Figure 10.11a shows that due to the low thermal conductivity, the thermal backfill made of the sand-cement mix does not ensure the flexibility to increase the current loading over 1,100 A and keeps the cable temperature below 65 °C. The thermal performance of the FTB backfill material is

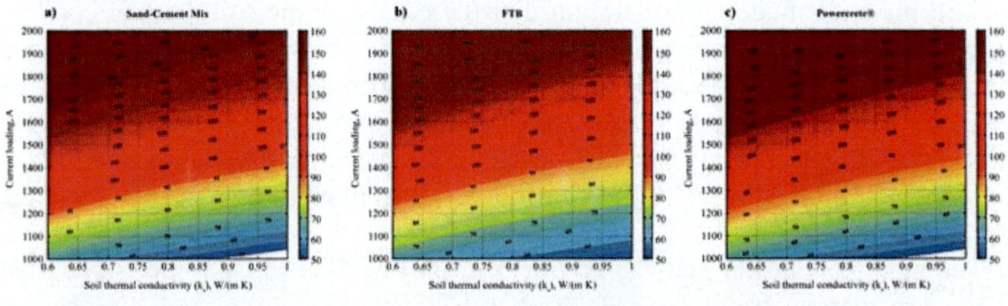


**Fig. 10.11** Cable conductor maximum temperature in °C determined for the design variables given in Table 10.9, the soil thermal conductivities ranging from 0.6 to 1.0 W/(m K) and the cable ampacity of 1000–2000 A for Different Thermal Backfills: **a** Sand-Cement Mix , **b** FTB and **c** POWERCRETE™

**Table 10.9** Values of the design variables considered in the thermal performance analysis of different thermal backfills

| Parameter | $l$, m | $p$, m | $b$, m | $s$, m | $A_c$, m$^2$ |
|---|---|---|---|---|---|
| Value | 0.4 | 0.3 | 0.3 | 0.2 | $1.6 \cdot 10^{-3}$ |

better compared to the sand-cement mix-based backfill and makes it possible to achieve allowable current loading of up to 1,165 A for $k_s$ = 1.0 W/(m·K). For the system under consideration, the best thermal/electrical performance is delivered by the POWERCRETE™ backfill by allowing an increase in the current loading that can be transferred safely without exceeding $T_{opt}$ to the level of 1,210 A for $k_s$ = 1.0 W/(m·K).

The cable ampacity that corresponds to the cable core temperature reaching 90 °C is equal to 1415 A for the sand-cement mix (Fig. 10.11a); 1,460A for FTB (Fig. 10.11b) and 1,530 A for POWERCRETE™ (Fig. 10.11 c). These are the values obtained for the soil thermal conductivity of $k_s$ = 1.0 W/(m·K). With a decrease in the soil thermal conductivity, the cable ampacity decreases at a constant temperature. Therefore for dry conditions (e.g. $k_s$ = 0.6 W/(m K)), the following values are obtained: 1200 A, for the sand-cement mix, 1240 A for FTB, and 1280 A for POWERCRETE™.

From the point of view of electricity distribution companies, the flexibility of the UTL is crucial because it allows local variations in current loading. Therefore, the recommended backfill is POWERCRETE™, as it allows the largest current loading variation to occur without exceeding the cable conductor allowable temperature.

### *10.5.4 Thermal Performance of the UPCS Under Various Soil and Backfill Thermal Conductivities*

Designing the UPCS, cable designers look for a thermal conductivity of the cable backfill material that ensures stable operating conditions of the UPCS in a wide range of changes in the soil thermal conductivity. The soil thermal conductivity exhibits considerable variations depending on the soil water content. In summer, the soil water content in the region of the cable line installation may decrease due to the combined effect of heat generation from the cable core and high ambient temperature. Also, plants growing at the ground level absorb moisture at the depth of up to 0.5 m. These two phenomena have an adverse impact on the soil thermal conductivity leading to its decrease. Therefore, the thermal backfill material needs to ensure safe operating conditions of power cables in a broad range of changes in the soil thermal conductivity. It should be noted that the cable backfill conductivity may decrease over time due to the material aging process. Therefore, the UPCS performance needs to be analysed for varied thermal conductivities of the cable backfill. Figure 10.12 shows the changes in the cable conductor maximum temperature with the thermal conductivity of the applied cable backfill and soil for different current loadings $I$ varying from 1000 A, through 1145 A, up to 1300 A. The cable backfill thermal conductivity ranges from 1.0 to 3.0 W(m·K) and the soil thermal conductivity—from 0.6 to 1.0 W/(m K). The backfill dimensions and the cable conductor diameter are considered the same as given in Table 10.9.

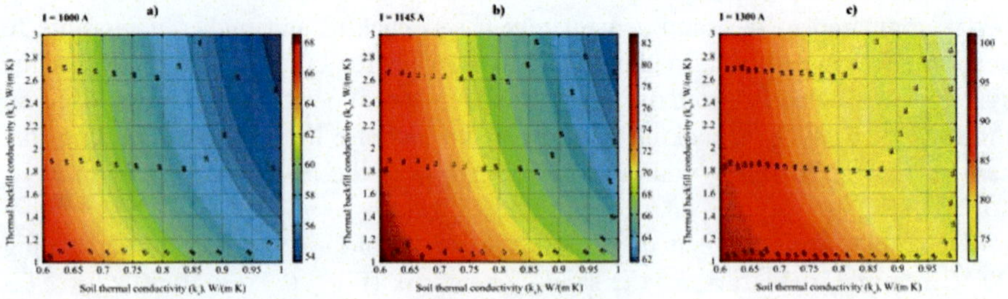


**Fig. 10.12** Cable conductor maximum temperature in °C determined for the design variables given in Table 10.9, the soil thermal conductivity ranging from 0.6 to 1.0 W/(m K) and the cable backfill thermal conductivities ranging from 1.0 to 3.0 W/(mK) for different current loadings: $I = 1{,}000$ A, $I = 1{,}145$ A and $I = 1{,}300$A

Figure 10.12a shows that for the current loading of $I = 1000$ A, the maximum temperature of the cable conductor is lower than the highest allowable temperature of the cable operation (90 °C) for the considered range of the soil and cable backfill thermal conductivities. However, as it can be seen in Fig. 10.12a, the backfill thermal conductivity below 2.4 W/(m K) does not make it possible to achieve the optimal temperature of the cable operation (65 °C) in the entire range of changes in the soil thermal conductivities under consideration.

For $I = 1145$ A (Fig. 10.12b), the optimal temperature of the cable conductor is achieved for $k_s = 1.0$ W/(m K) and $k_b > 1.4$ W/(m K). However, for the current loading of $I = 1145$ A, the cable conductor temperature is lower than the critical temperature of the cable operation (90 °C) within the entire range of variations in the soil and cable backfill thermal conductivities under consideration.

When the current loading is increased to 1300 A (Fig. 10.12c), the optimal temperature of the cable operation (65 °C) is not possible for the entire range of $k_s$ and $k_b$ variations. In this case, the cable conductor temperature is below the critical value (90 °C) for the soil thermal conductivity higher than 0.75 W/(m K) for the considered range of the cable backfill thermal conductivity.

## 10.6 Outline

This chapter presents the application of a modified advanced optimization algorithm named as modified Jaya algorithm for optimizing the installation (material) costs of the underground power cable system (UPCS). The installation costs included the costs of cables and the cable backfill material. As an example, the performance optimization of a 400 kV high-voltage underground cable line is presented. The cable system operating in a flat (in-line) formation is considered. The cable burial depth is constant and equal to 2.0 m while the thermal backfill layer where the cable line is placed has varied dimensions to ensure the optimal thermal performance of the cable system. The practical objective of this study is to design the cable system in such a way that:

- the material costs (cable costs + thermal backfill costs) are minimized,
- the maximum temperature of the cable conductor is equal to the optimum temperature for the considered current loading (in this optimization study the current loading and the cable conductor optimum temperature are adopted as 1145 A and 65 °C, respectively).

Thermal backfills that improve heat removal from underground power cables are in particular cases used by cable engineers for high-voltage power cables, mostly when the risk of the cable overheating is anticipated. The thermal performance of three commonly used backfill materials is studied:

- Sand-cement mix (with proportions of 12:1, thermal conductivity of 1.0 W/(m K) and unit costs of 20 Euro per cubic meter),
- Fluidized Thermal Backfill™ with thermal conductivity of 1.54 W/(m K) and unit costs of 125 Euro per cubic meter,
- POWERCRETE™, developed by the Heidelberg Cement Group, with the thermal conductivity of 3.0 W/(m K) and approximated unit costs of 400 Euro per cubic meter.

The thermal conductivity values considered for the thermal backfill materials are assumed for the dry state. Such conditions may occur in the vicinity of the cable line where the cable temperature is the highest and the liquid vapour transport occurs. The assumption of the thermal backfill dry state ensures the safety margin in the analysis of the backfill thermal performance (the cable conductor temperature for the model used in the simulation is higher than for the real object).

This chapter also investigates the influence of the soil thermal conductivity on the thermal performance and material costs of the cable system. The optimization performed for varied soil thermal conductivities and thermal backfill materials made it possible to determine the cable backfill layer dimensions, the cable spacing and the cable diameter in a way that enables minimization of material costs.

The model of the heat transfer in the UPCS is considered as a steady-state, two-dimensional heat conduction problem with the source term. The finite element method is used to discretize the heat conduction equation and to determine the temperature distribution within the entire UPCS. The maximum temperature of the cable conductor is obtained based on the electric-thermal model of the UPCS output.

The optimization is performed using the modified Jaya algorithm, which in the evolutionary equation uses randomly selected three best and three worst solutions and updates the set of design variables in consecutive iterations until convergence is achieved (i.e. all solutions in the population are the same). In the proposed optimization procedure using the modified Jaya algorithm, a static penalty function is used to check the constraint on the optimum value of the cable conductor temperature. The penalty function value is calculated based on the difference between the cable conductor temperature determined from the UPCS model and the optimum value of the cable operating temperature specified by the cable producer. The performance of the modified Jaya algorithm is compared with the classical Jaya algorithm and the Particle Swarm Optimization (PSO) algorithm. The convergence behaviour of the modified Jaya algorithm is tested for different population sizes, and the algorithm achieves the lowest value of the cost function, compared to the classical Jaya algorithm and the PSO algorithm for the analysed case.

The following conclusions are derived based on the thermal model of the UPCS output:

- For the soil thermal conductivity varied in the range of 0.6–1.0 W/(m K), and in-line formation of the 400 kV UPCS, the sand-cement mix performs poorly if the thermal conductivity value is lower than 0.75 W/(m K). For the case under consideration, it is impossible to design a cable system where the cable conductor temperature is kept below 65 °C.
- The POWERCETE™ thermal backfill made it possible to achieve the optimum temperature of the cable conductor for the cases considered during the design optimization of the UPCS.
- The FTB backfill is a cheaper alternative to POWERCRETE™ and in particular cases makes it possible to achieve lower material costs than the POWERCRETE™ backfill.
- Despite the high costs of the advanced backfill materials (POWERCRETE™ and FTB), the overall material costs of the UPCS are the highest when the sand-cement mix is applied. This is because both the POWERCRETE™ and the FTB backfills make it possible to reduce the cable conductor cross-sectional area compared to the sand-cement mix. From the point of view of electricity distributors, the profits from the reductions in the cable cross-sectional area are significantly higher than the losses due to the application of expensive backfill materials. The reduction in material costs is up to 200 kEuro for the case with the soil thermal conductivity value of 1.0 W/(m·K).
- The POWERCRETE™ backfill can be regarded as the best of the considered backfill materials if high flexibility of the cable line is expected. The comparative study of the thermal/electrical performance of different backfill materials is carried out for the thermal bedding width and height of 1.2 and 0.6 m, respectively. A rectangular shape of the thermal backfill layer is considered. When compared to the commonly used sand-cement mix for the same cross-sectional area of the backfill layer, the application of the POWERCRETE™ backfill allows an increase in the cable maximum ampacity (in the critical temperature of 90 °C) of up to 115 A. The FTB backfill allows an increment in the current loading of up to 45 A.

## References

Adams JI, Baljet AF (1968) Thermal behaviour of cable backfill materials. IEEE Trans Power Apparatus Syst 87:1149–1161

An EU Strategy on Heating and Cooling (2016) COM, Final Europ Commission: Brussels Belgium 51:2016
Al-Saud M, El-Kady M, Findlay R (2007) A novel finite-element optimization algorithm with applications to power cable thermal circuit design. In: Power engineering society general meeting, IEEE, pp 1–8
Campbell G, Jungbauer J Jr, Bidlake W, Hungerford R (1994) Predicting the effect of temperature on soil thermal conductivity. Soil Sci 158(5):307–313
Cuvelier F, Japhet C, Sarella G (2013) An efficient way to perform the assembly of finite element matrices in Matlab and Octave. Technical Report. 8305, Universite de Paris
Davis M, Maples T, Rosen B (2010) Cost-saving approaches to wind farm design. TEChBriefs 4:1e5
Lieto Vollaro De, Fontana L, Vallati A (2011b) Thermal analysis of underground electrical power cables buried in non-homogeneous soils. Appl Therm Eng 31(5):772–778
De Lieto Vollaro R, Vallati A (2013) Study of a model for the evaluation of the heat losses from electric cables buried according to the norm standard. Adv Mater Res 650:437–442
De Lieto Vollaro R, Fontana L, Quintino A, Vallati A (2011a) Improving evaluation of the heat losses from arrays of pipes or electric cables buried in homogeneous soil. Appl Therm Eng 31(17–18):3768–3773
Del Pino-López JC, Cruz-Romero P, Serrano-Iribarnegaray L, Martínez-Román J (2014) Magnetic field shielding optimization in underground power cable duct banks. Electr Power Syst Res 114:21–27
ENERGETIK DK (2009) Cable action plan 132–150 kV grids
Electric Power Research Institute Inc. (2003) Increased power flow guidebook—underground cables. Final Report
Gouda OE, El Dein AZ, Amer MG (2011) Effect of the formation of the dry zone around underground power cables on their rating. IET Proc. Trans Power Deliver 26:972–978
Galli G, Vallati A (2011) Equilibrium heat and moisture transfer in soils with heat sources. Int J Heat Technol 29(1):33–38
Hiivala LJ, Landinger CC (2014) Conductors. In: Chapter in electrical power cable engineering. 3rd Edn. CRC Press
Hwang CC, Jiang YH (2003) Extensions to the finite element method for thermal analysis of underground cable systems. Electri Power Syst Res 64(2):159–164
IEC 60287-2-1 (1995) Electric cables—calculation of the current rating—part 2: thermal resistance—section 1: calculation of the thermal resistance
Kennedy J, Eberhart RC (1995) Particle swarm optimization. Proceed IEEE Int Conf Neural Netw 4:1942–1948
Kennedy J, Eberhart RC, Shi Y (2001) In: Swarm intelligence. Morgan Kaufmann
Kovač N, Sarajčev I, Poljak D (2006) A numerical–stochastic technique for underground cable system design. IEEE Proceed-Gener Transm Distrib 153(2)
Kroener E, Vallati A, Bittelli M (2014) Numerical simulation of coupled heat, liquid water and water vapor in soils for heat dissipation of underground electrical power cables. Appl Therm Eng 70(1):510–523
Metwally I, Al-Badi A, Al Farsi A (2013) Factors influencing ampacity and temperature of underground power cables. Electri Eng 95(4):383–392
Moutassem W, Anders GJ (2010) Configuration optimization of underground cables for best ampacity. Power Delivery IEEE Trans 25(4):2037–2045
NRG Expert, Electricity T&D White Paper (2013). nrgexpert.com
Ocłoń P, Cisek P, Taler D, Pilarczyk M, Szwarc T (2015a) Optimizing of the underground power cable bedding using momentum-type particle swarm optimization method. Energy 92(2):230–239
Ocłoń P, Cisek P, Rerak M, Taler D, Rao RV, Vallati A, Pilarczyk M (2018) Thermal performance optimization of the underground power cable system by using a modified Jaya algorithm. Int J Therm Sci 123:162–180

Ocłoń P, Cisek P, Pilarczyk M, Taler D (2015b) Numerical simulation of heat dissipation processes in underground power cable system situated in thermal backfill and buried in a multilayered soil. Energy Convers Manage 95:352–370

Ocłoń P, Rerak M, Rao RV, Cisek P, Vallati A, Jakubek D, Rozegnał B (2021) Multiobjective optimization of underground power cable systems. Energy 215(119089)

Ocłoń P, Taler D, Cisek P, Pilarczyk M (2015) Fem-based thermal analysis of underground power cables located in backfills made of different materials. Strength Mater 47(5):770–780

Papagiannopoulos I, Chatziathanasiou V, Exizidis L, Andreou G, De Mey G, Wiecek B (2013) Behaviour of the thermal impedance of buried power cables. Int J Electr Power Energy Syst 44(1):383–387

Popiel CO, Wojtkowiak J, Biernacka B (2001) Measurements of temperature distribution in ground. Exp Therm Fluid Sci 25:301–309

Rao RV (2016) Jaya: A simple and new optimization algorithm for solving constrained and unconstrained optimization problems. Int J Indus Eng Comput 7:19–34

Radhakrishna HS (1982) Fluidized cable thermal backfill. In: Boggs SA et al. (ed) Underground cable thermal backfill. Pergamon Press, pp 34–53

Salata F, Nardecchia F, de Lieto Vollaro A, Gugliermetti F (2015) Underground electric cables a correct evaluation of the soil thermal resistance. Appl Therm Eng 78(0):268–277

TELE-FONIKA Company (2012) In: High voltage cables. 1st edn cracow (in polish)

Thue WA (2011) In: Electrical power cable engineering. 3rd Edn. CRC Press, pp 307

Valle YD, Venayagamoorthy GK, Mohagheghi S, Hernandez JC, Harley RG (2008) Particle swarm optimization: basic concepts variants and applications in power systems. Evolut Comput IEEE Trans 12(2):171–195

Zarchi DA, Vahidi B, Haji MM (2014) Optimal configuration of underground cables to maximise total ampacity considering current harmonics. Gener Transm Distrib IET 8(6):1090–1097

# Summary and Conclusions

This monograph discusses underground energy systems. In the first part, the RESHeat system, a novel solution for heating residential buildings is presented. The second part presents optimization of underground cable systems using the modified Jaya algorithm.

The RESHeat system is put forward as a novel solution for heating buildings. The system is based on renewable energy sources. A water-to-water ground source heat pump is used as the heat source. The system operation is assisted by PVT cells and solar collectors with a sun-tracking system. In addition, a ground heat accumulation unit is proposed in the form of two tanks: a heat accumulator (a non-insulated reservoir exchanging heat with the ground) and an insulated storage tank (used for domestic hot water production and as the heat pump lower heat source if the amount of heat in the non-insulated reservoir is insufficient). The monograph presents the COP testing results of a ground source heat pump in a demonstration installation of the RESHeat system at the headquarters of the *F.H.U. Urządzenia Chłodnicze Czamara* company in Limanowa (Poland). The system mathematical model is also developed, which makes it possible to analyse the degree of coverage of the demand for thermal energy by using accumulation tanks as the heat pump lower source.

An analysis is also conducted of the impact of the accumulation tank diameter on the efficiency of the heat exchange with the ground. It is demonstrated that by decreasing the diameter from 3 to 2 m, the degree of the RESHeat system coverage of the building demand for thermal energy is reduced by 11%. Optimization is also performed of the ground source heat pump heat extraction from the accumulation unit for two different diameters of the tank, i.e. for 2 and 3 m. A Particle Swarm Optimization algorithm is used. The analysis also covers the impact of the ground physical properties (heat conductivity coefficient, density, specific heat) on the average seasonal COP of the heat pump and on the degree of coverage of the demand for heat by the RESHeat system.

The key element of the system are PVT (photovoltaic thermal) cells that produce both electricity and heat. The waste heat from the PVT cells is stored in an accumulation tank buried in the ground. The monograph presents a mathematical model



P. Ocłoń, *Renewable Energy Utilization Using Underground Energy Systems*, Lecture Notes in Energy 84, https://doi.org/10.1007/978-3-030-75228-6

of the PV cells cooling system—a radiator using U-tube segments. An analysis is conducted of the effect of the number of segments on the heat transfer efficiency for the applied cooling system. The developed model makes use of the finite volume method—a three-dimensional equation of heat conduction within the PV panel and the cooling system and a one-dimensional energy balance equation for the cooling medium (water–glycol mixture). It is demonstrated that the cooling is the most effective in the case of 6 segments, and this is the system implemented by the ELFRAN company. The mathematical model of the PV cells cooling system is verified experimentally on a test stand made within the HySOL project (financed by the NCR&D). The experimental tests carried out over a period of one day showed satisfactory agreement between the results obtained from the model and from the measurements.

The monograph also presents a cost-effectiveness analysis of the RESHeat-based investment intended for building heating. The economic analysis is performed for two buildings with 20 and 50 apartments, respectively. The buildings (constructed in the year 1970) demonstrate a poor energy standard. It is proved that the investment outlays on the RESHeat system, due to the system minimal maintenance costs, are characterized by the payback period of 4–8 years. The shortest payback period was obtained by comparing the RESHeat system with heating with a ground source heat pump (3.5 years for the building with 20 apartments and 4 years for the building with 50 apartments). The investment longest payback period is obtained for the comparison of the RESHeat system with heating using a condensing boiler (8 years for the building with 20 apartments) and with district heating (8.5 years for the building with 50 apartments).

The RESHeat solution can be integrated with a variety of existing heat sources, including biomass or natural gas boilers and district heating systems. In the case of natural gas or biomass boilers, the RESHeat system will use them as the peak heat source. If there is excess demand for heat that the heat pump is unable to meet, a natural gas or biomass boiler can be used to raise the temperature in a buffer tank from which heat is transferred to fan-coil units, radiators or to the underfloor heating system.

If the building is connected to the district heating network, the system of the heat pump upper source can be connected to the building heating main and thus reduce the heat consumption of the district heating network.

Another important aspect of the potential application of the RESHeat solution is that the system can be redesigned to minimize the space occupied by the underground accumulation unit. Horizontal storage tanks can be replaced with vertical ones, which take up less space required for the installation. The disadvantage of vertical storage tanks is the non-uniform temperature of water, which affects the performance of the storage unit. However, if the RESHeat system is designed correctly, the effect of the water temperature non-uniformity is marginal.

The RESHeat system can also act as an alternative to existing solutions. If a natural gas boiler needs to be replaced (in the case of failure), it is possible to connect the RESHeat system to the existing heating system and provide heating through a radiator or underfloor heating. The RESHeat solution can thus provide a fully renewable energy system and be combined with the existing heating system.

Nowadays, with the ever-increasing share of renewables, high-voltage underground power cable systems attract significant attention. Underground cable lines are prone to failure due to the excessive heat emitted from the cables. This refers especially to high-voltage underground power cables. When the soil is dry, the severe heat dissipation conditions may lead to the cable overheating and costly failures. Therefore, there is a need to apply cable backfill materials to create appropriate conditions for more efficient heat dissipation from cables and to protect the cables from overheating. This monograph proposes an optimization algorithm—a modified Jaya method—which is used to design underground power cable systems. The algorithm selects the most cost-effective backfill material to ensure that the cable conductor temperature does not exceed 65 °C (the design temperature). It is found that the POWERCRETE™ backfill manufactured by the Heidelberg Cement Group is a very efficient solution to prevent the overheating of high-voltage underground power cables.